HEALTHY BABIES ARE WORTH THE WAIT

HEALTHY BABIES ARE WORTH THE WAIT

Edited by
EDWARD R.B. McCABE

Amsterdam • Boston • Heidelberg • London
New York • Oxford • Paris • San Diego
San Francisco • Singapore • Sydney • Tokyo
Academic Press is an imprint of Elsevier

Academic Press is an imprint of Elsevier
125 London Wall, London EC2Y 5AS, UK
525 B Street, Suite 1800, San Diego, CA 92101-4495, USA
225 Wyman Street, Waltham, MA 02451, USA
The Boulevard, Langford Lane, Kidlington, Oxford OX5 1GB, UK

Library of Congress Cataloging-in-Publication Data
A catalog record for this book is available from the Library of Congress

British Library Cataloguing-in-Publication Data
A catalogue record for this book is available from the British Library

ISBN: 978-0-12-803482-8

For information on all Academic Press publications
visit our website at http://store.elsevier.com/

Typeset by TNQ Books and Journals
www.tnq.co.in

Printed and bound in the United States of America

CONTENTS

AUTHORS

PRIMARY AUTHORS

Diane M. Ashton
March of Dimes Foundation, National Office, White Plains, NY, USA and Department of Obstetrics & Gynecology, SUNY Downstate Medical Center, Brooklyn, NY, USA

Karla Damus
March of Dimes Foundation, National Office, White Plains, NY, USA (at commencement of study); Department of Family Medicine, Boston University School of Medicine/Boston Medical Center, Boston, MA, USA (current)

CONTRIBUTING AUTHORS

Vani R. Bettegowda
Perinatal Data Center, March of Dimes Foundation, National Office, White Plains, NY, USA (at time of study)

Gerard E. Carrino
Program Research Development and Evaluation, March of Dimes Foundation, National Office, White Plains, NY, USA

Todd Dias
Perinatal Data Center, Program Research Development and Evaluation, March of Dimes Foundation, National Office, White Plains, NY, USA

Tracey Jewell
Division of Maternal & Child Health, Kentucky Department for Public Health, Frankfort, KY, USA

Joy Marini
Maternal, Newborn and Child Health, Johnson & Johnson Corporate Contributions, New Brunswick, NJ, USA

Rebecca Russell
Perinatal Data Center, Program Research Development and Evaluation, March of Dimes Foundation, National Office, White Plains, NY, USA

Ruth Ann Shepherd
Maternal and Child Health, Kentucky Department for Public Health, Cabinet for Health and Family Services, Frankfort, KY, USA

Julie Solomon
J. Solomon Consulting, LLC, Mountain View, CA, USA

CONTRIBUTORS FOR DATA ANALYSIS OF THE *HEALTHY BABIES ARE WORTH THE WAIT*® CONSUMER AND PROVIDER SURVEYS

Lorena Ortiz
J. Solomon Consulting, LLC, Mountain View, CA, USA

Monika Sawhney
J. Solomon Consulting, LLC, Mountain View, CA, USA

EDITOR

Edward R.B. McCabe
March of Dimes Foundation, National Office, White Plains, NY, USA

FOREWORD

HEALTHY BABIES ARE WORTH THE WAIT®: REDUCING PRETERM BIRTH THROUGH THE INTEGRATION OF PRIMARY CARE AND PUBLIC HEALTH

The Challenge

In a report of births occurring in the US during the year 2006, it was stated, "Preterm and low birthweight rates continued to rise" [1]. This pessimistic statement, suggesting a rise that could not be altered, was justified by more than two decades of steadily increasing rates of preterm birth or births less than 37 weeks of gestation. The preterm birth rate had risen by more than 20% from 1990 and 36% from the early 1980s. This increase in preterm birth was influenced not as much by births before 34 weeks, which rose "modestly" from 3.32% in 1990 to 3.66% in 2006, and more by late preterm birth (34–36 weeks) which increased by 25% from 7.3% in 1990 to 9.15% in 2006. This same report noted that efforts to effectively prevent preterm birth were "limited," even though over one-third of infant deaths were thought to be related to late preterm birth, and infants born late preterm had increased risks of long-term consequences.

Fortunately, not everyone considered the prevention of preterm birth an intractable problem, including the March of Dimes Foundation, which, supported by its Board of Trustees, declared preterm birth prevention to be a priority for the Foundation in 2003, ultimately revising the March of Dimes mission in 2005 to include the problem of premature birth in the US as a major focus and expanding this to a global campaign in 2008. Through the hard work of many organizations and individuals the preterm birth rate dropped in 2012 for the sixth straight year since the peak in 2006 to 11.5%, the lowest in a decade [2].

Notably, the Association of State and Territorial Health Officials (ASTHO) created the Healthy Babies President's Challenge of 2012 led by Dr David Lakey, Texas Commissioner of Health. This led to a committed pledge from the 50 states, Washington, DC, and Puerto Rico to reduce preterm births by 8% by 2014 [3].

Mobilizing an Integration of Primary Care and Public Health

The efforts of these individuals and organizations to reduce preterm birth have included programs involving the integration of primary care and public health as described in the consensus report from the Institute of Medicine (IOM), *Primary Care and Public Health: Exploring Integration to Improve Population Health* [4], and the ASTHO Presidential Challenge by ASTHO President and New Hampshire Director of Public Health Services, Jose Montero, MD, To Advance the Reintegration of Public Health and Health Care [5], both in 2012.

The IOM report [4] describes a continuum of integration between primary care and public health from isolation to merge with the goal being to achieve more efficacious improvement of the health of individuals, their communities, and the entire population. The intermediate steps along this continuum of integration include: mutual awareness—informing each other of their activities; cooperation—sharing resources between those in primary care and public health; collaboration—planning and executing with more intense and joint activity between these sectors; and partnership—integrating the activities of the two sectors programmatically so that they appear inseparable. The report recognizes that initiatives may not be at the partnership or merger levels of integration at the outset, but their position within this continuum should not be a barrier to their initiation; the understanding of the steps along the way will provide a roadmap for improving integration in the process. The report boldly states, "Beginning is more important than waiting until all of the requisite components are in place." Key principles in this process include strong leadership to foster integration across perceived disciplinary, programmatic, and jurisdictional boundaries; sustainability to maintain continuing success; and collaboration in the acquisition and analysis of data.

ASTHO President, Dr Montero, challenged his colleagues to use systems approaches to "implement integrated efforts that improve population health and lower health cost" [5]. In addition to Primary Care and Public Health [4], he referenced the IOM reports, For the Public's Health: Investing in a Healthier Future [6] and Best Care at Lower Cost: The Path to Continuously Learning in Health Care in America [7]. This ASTHO challenge built on the past ASTHO President's Challenge focused on Health Equity and Dr Lakey's Challenge on Healthy Babies. As state health officers directing the front lines of the public's health, it is appropriate that cost effectiveness, as well as efficacy, be considered and these efficiencies can be best achieved through collaborative practices that enhance services and reduce duplication of effort. To implement these integrated efforts, ASTHO worked in collaboration with the United Health Foundation, the Institute of Medicine and over 50 leading medical and public health associations in the ASTHO-supported

Primary Care Public Health Collaboration [8]. These national partners have generated a plan to develop and demonstrate successes, to realign resources to coordinate and to sustain these initiatives, to disseminate and bring to scale effective initiatives, and to design effective metrics and infrastructure for population health.

Building on these successful initiatives, ASTHO President Terry Cline, PhD, Health Commissioner of Oklahoma, issued his *Presidential Challenge, Reducing Harms Associated with Prescription Drugs* to curb prescription drug misuse and abuse [9]. This is intended to address an issue of concern to the March of Dimes and ASTHO, the increasing incidence of neonatal abstinence syndrome.

Examples of Integration and Evaluation Strategies to Reduce Preterm Birth

A 2012 article in *The Lancet* described the impact that bundling evidence-based strategies with demonstrated effectiveness would have on preterm birth in 39 countries with very high human development index (VHHDI) [10]. To be effective, these approaches would require the integration of activities among public health agencies, hospitals, community health clinics, practices, and health-care providers. These integrated strategies included smoking cessation, reducing the transfer of multiple embryos with assisted reproductive technology, cervical cerclage, progesterone therapy, and decreasing nonmedically indicated deliveries by induction or caesarean birth. The estimated impact of these interventions for the 39 VHHDI countries was a 5% reduction in preterm birth, and would represent 58,000 fewer preterm births annually and an estimated total cost savings of approximately $3 billion US dollars in direct and societal costs. Therefore, integration of activities across delivery platforms and providers involving primary care and public health, community advocates, and engaged community participation, could be extremely effective in reducing preterm birth with the knowledge we have available today.

In Texas, Dr Lakey reports that the state has effectively integrated systems to address high rates of preterm birth and infant mortality. The Texas Department of State Health Services successfully worked with 11 local coalitions in 23 counties with high Medicaid patient populations. This was coordinated with a state policy change to stop Medicaid payment for non-medically necessary inductions before 39 weeks. The integrated effort included regional work with providers to improve access to care and improved health services for all women.

Summary of the *Healthy Babies are Worth the Wait®* Initiative

The initiative described in this book demonstrates the power of integrating primary care and public health. The Kentucky Department for Public Health, in collaboration with the March of Dimes Kentucky Chapter and with resources and direction from the March of Dimes Foundation and the Johnson & Johnson Pediatric Institute, LLC participated in the implementation of a pilot program to reduce preterm birth in three selected hospital community sites in Kentucky. This community-based initiative built upon existing and created new partnerships among public health professionals, hospital clinicians, and community advocates, and required experts and leaders, such as Dr William Hacker, former Kentucky Health Commissioner and Dr Ruth Ann Shepherd, Kentucky Maternal Child Health Director, with knowledge of local health-care delivery systems, providers, and patients to provide guidance in the development of culturally sensitive and health literacy appropriate educational materials. In *Healthy Babies are Worth the Wait®* communities, leaders and staff of the local hospitals and local health departments met together regularly, studied best practices in both clinical and public health services, worked to build a system of services with enhanced referral processes and removal of barriers, and developed consistent messaging that included public messaging, so that pregnant women received the same information at every touch point in their care, as well as from their communities. This team approach tailored to local needs resulted in a significant reduction of the preterm birth rate in the intervention sites that employed *HBWW*. Lesser reductions were also observed in the comparison sites and the rest of Kentucky; during the project period and as the sites showed success, there was an intentional spread of information and best practices about *HBWW* in a state-wide effort to decrease the burden of preterm birth. An important lesson from this pilot program is that partnerships between primary care providers, public health professionals, community advocates, and the local community can achieve a desired goal for their community's health, in this case a reduction in preterm birth.

Integrating Primary Care and Public Health to Reduce Preterm Birth

Preterm birth is a major health problem in the US and internationally with serious consequences in terms of mortality and lifelong morbidity for survivors. The human consequences are staggering and the economic burdens are tremendous. Preterm birth was once thought to be a problem so complex as to be hopelessly intractable. However, through the efforts of the March of Dimes and others, the preterm birth rate in the US has been declining since its peak in 2006. The IOM report, *Primary Care and Public Health: Exploring Integration to Improve Population Health* [4], and the ASTHO Presidential Challenge by Dr Montero [5] describe the need to integrate the primary care and public health sectors to address the problems facing the population's health in addition to mobilizing the committed engagement of the public, communities, and stakeholders at all levels. Preterm birth is one of these serious population health problems that demand this integration. The *HBWW* pilot initiative described in this book illustrates an example of successful integration of primary care, public health, community advocates, and community engagement to reduce preterm birth in Kentucky. Based on the success of this pilot, the March of Dimes is expanding this program to offer it to additional communities in Kentucky, New Jersey, and Texas, and plans to make it more broadly available nationally.

Authors

Edward R.B. McCabe, MD, PhD
March of Dimes Foundation, White Plains, NY, USA

Paul Jarris, MD, MBA
Association of State and Territorial Health Officials, Arlington, VA, USA

Jennifer L. Howse, PhD
March of Dimes Foundation, White Plains, NY, USA

REFERENCES

[1] Martin JA, Hamilton BE, Sutton PD, Ventura SJ, Menacker F, Kirmeyer S, et al. Births: final data for 2006. Natl Vital Stat Rep 2009;57:1–104.
[2] March of Dimes. Premature birth report cards. 2013. Data available prior to release on November 1, 2013.
[3] ASTHO. Healthy babies. Available at: http://www.astho.org/healthybabies/ [accessed 19.10.13].
[4] Committee on Integrating Primary Care and Public Health, Board on Population Health and Public Health Practice, Institute of Medicine. Primary care and public health: exploring integration to improve population health. Washington, DC The National Academies Press; 2012. Available at: http://www.iom.edu/Reports/2012/Primary-Care-and-Public-Health.aspx [accessed 19.10.13].
[5] Montero JT. To advance the reintegration of public health and health care; ASTHO. Available at: http://www.astho.org/WorkArea/DownloadAsset.aspx?id=7479 [accessed 19.10.13].
[6] Committee on Public Health Strategies to Improve Health, Institute of Medicine. For the public's health: investing in a healthier future. Washington, DC: The National Academies Press; 2012. Available at: http://books.nap.edu/openbook.php?record_id=13268 [accessed 19.10.13].

[7] Committee on the Learning Health Care System in America, Institute of Medicine. Best care at lower cost: the path to continuously learning health care in America. Washington, DC: The National Academies Press; 2012. Available at: http://books.nap.edu/openbook.php?record_id=13444 [accessed 19.10.13].

[8] ASTHO-Supported Primary Care and Public Health Collaborative. Available at: http://www.astho.org/Programs/Access/Primary-Care-and-Public-Health-Integration/ [accessed 10.11.13].

[9] ASTHO. Presidential challenge – reducing harms associated with prescription drugs. 2014. Available at: http://www.astho.org/rx/ [accessed 10.11.13].

[10] Chang HH, Larson J, Blencowe H, Spong CY, Howson CP, Cairns-Smith S, et al. Preventing preterm births: analysis of trends and potential reductions with interventions in 39 countries with very high human development index. Lancet 2013;381:223–34.

ACKNOWLEDGMENTS

This project was jointly funded by Johnson & Johnson Pediatric Institute, LLC and March of Dimes Foundation. The authors thank Dr Jennifer L. Howse, President of the March of Dimes Foundation, for her visionary leadership and unwavering investment in support of the *Healthy Babies are Worth the Wait*® (*HBWW*) initiative as a novel intervention to combat the escalating problem of preterm birth in the United States. We are also grateful to Bonnie Petrauskas, Executive Director, Corporate Contributions & Community Relations and Joy Marini, MS, PA-C, MBA, Executive Director, Corporate Contributions Maternal, Newborn and Child Health at Johnson & Johnson for their leadership and valuable contributions as a strategic thought partner in the conceptualization and deployment of this initiative. The longstanding relationship established between Johnson & Johnson Pediatric Institute, LLC and the March of Dimes was foundational to developing the *HBWW* initiative. We acknowledge Dr William Hacker, former Health Commissioner of the Kentucky Department for Public Health and Dr Ruth Ann Shepherd, Division Director for Kentucky Maternal and Child Health for providing exceptional leadership as dedicated partners during the *HBWW* initiative in Kentucky.

We are deeply indebted to Dr Karla Damus whose profound expertise in maternal and child health guided the concept and design for *HBWW* and skillfully led the initiative throughout its implementation. Many thanks to Katrina Smith, Director of Program Services and her staff at the March of Dimes Greater Kentucky Chapter, the *HBWW* teams at the intervention and comparison site hospitals, the local health departments at the intervention sites, the *HBWW* Advisory Committee, and the patients and community members who participated in the initiative and were instrumental to its ultimate success in demonstrating a reduction in preterm birth rates.

CHAPTER 1

Introduction

The March of Dimes Foundation and Johnson & Johnson Pediatric Institute, LLC in collaboration with the Kentucky Department for Public Health (KYDPH), began the planning phase of a preterm birth (less than 37 completed weeks of gestation) prevention initiative entitled *Healthy Babies are Worth the Wait*® (*HBWW*) in 2006. The project was designed by Dr Karla Damus, a consulting senior epidemiologist at the March of Dimes, as a 3-year (2007–2009) multifaceted initiative to prevent "preventable" preterm births. The cornerstone of the initiative was the partnerships among hospital clinicians, public health professionals, and community advocates, complemented by extensive educational outreach to the community. This partnership facilitated the implementation of patient-, provider-, and community-level interventions to address multiple modifiable risk factors for preterm birth and ultimately improve patient services to reduce preterm births. The initiative was developed to test if clinical and public health collaborations implemented through bundled health-care delivery interventions could decrease preterm births in selected communities in Kentucky.

CHAPTER 2

Background

In 2006, a year prior to the launch of *Healthy Babies are Worth the Wait*® *(HBWW)* in Kentucky, the US preterm birth rate reached an all-time high of 12.8%, reflecting a 30% increase over the previous three decades [1]. Preterm infants are at increased risk for newborn complications such as respiratory distress syndrome; intraventricular hemorrhage, infections, and apnea; and for life-long impediments that include intellectual and developmental disabilities, cerebral palsy, blindness, and physical and neurological impairment [2,3]. Preterm birth is also the leading cause of neonatal mortality—in 2006, 36% of infant deaths in the US were due to preterm-related causes [4]. During that same year, the Institute of Medicine reported that preterm births cost an estimated $26 billion each year in maternal and infant medical care, educational support, and lost productivity [3].

The rise in preterm birth is primarily attributable to the escalating rates of late preterm birth (occurring between 34 0/7 and 36 6/7 weeks gestation), which constituted more than 70% of all preterm births and accounted for approximately 390,000 births in 2006 [1]. Late preterm infants represent the largest proportion of neonates admitted into the neonatal intensive care unit (NICU) and are responsible for most NICU costs [5]. Compared to term infants, late preterm infants are also much more likely to be rehospitalized in the first year of life. They are 1.7 times more likely to be rehospitalized during the neonatal period secondary to complications of hyperbilirubinemia, feeding difficulties, and suspected sepsis [6]; are three times more likely to die before their first birthday; and have increased neonatal complications with jaundice, feeding difficulties, hypoglycemia [7], and long-term learning and behavioral problems [8]. Ultimately, success in reducing preterm birth, and in particular late preterm birth, will have a profound impact on the health-care system, associated costs, the health of individuals, and the well-being of families.

Most early preterm births and a substantial proportion of late preterm births are spontaneous and occur as a result of multiple pathways, including complex interactions of genetic, environmental, social, and behavioral factors [2,9]. Among the well-accepted risk factors for preterm birth are a history of preterm labor and delivery; multiple gestation pregnancy; infection and inflammation; preeclampsia; late or inadequate prenatal care; maternal cigarette smoking, alcohol consumption, and other substance use; and stress [2,9].

In addition, although medical intervention to deliver early often reflects optimal management, such as in cases of maternal and fetal complications, early delivery for nonmedical reasons has increased [10,11]. Providers are influenced by many factors, including a more litigious environment, and increasingly some pregnant women as consumers of health-care services request the mode and even the date of their delivery. These factors contribute to elevated elective induction and cesarean birth rates [12,13], which in turn are associated with higher preterm birth rates, especially those preterm births occurring in the late preterm period [14,15].

A number of risk factors for preterm birth, such as maternal cigarette smoking, alcohol consumption, other substance use, stress, and elective inductions or cesarean births at less than 39 weeks that may result in preterm birth due to inaccurate gestational dating, are potentially modifiable through interventions with pregnant women and perinatal care providers. One theory of change that underlies the *HBWW* interventions hypothesizes that positive changes in consumer and provider knowledge and attitudes relevant to these risk factors, in combination with the provision of educational resources for providers and consumer accessibility to relevant medical and psychosocial support services, have great potential to bring about positive changes that favor decreased preterm birth rates.

CHAPTER 3

Healthy Babies are Worth the Wait® Design and Methods

INITIATIVE DESIGN

Launched in March 2007, the *Healthy Babies are Worth the Wait®* *(HBWW)* initiative in Kentucky was designed as a 3-year (2007–2009 for the implementation), multifaceted initiative to decrease "preventable" preterm births by employing evidence-based clinical and public health interventions that would have a high likelihood of success during the project time period. The initiative utilized a mixed ecologic design, combining basic features of the multiple-group study and the time-trend study with the objective of making two types of comparisons simultaneously: change over time within intervention and comparison groups, and differences across groups [16,17].

The primary goal of *HBWW* in Kentucky was to reduce the local rate of singleton preterm birth by 15% over a 3-year period in three targeted intervention sites (IS) while observing contemporaneous outcomes in three comparison sites (CS), where pregnant women would receive conventional perinatal services. Intermediate objectives included reducing unnecessary obstetrical interventions, enhancing consumer education, and promoting healthy behaviors among pregnant women and families (Table 1). The *HBWW* logic model, illustrated in Appendix A, provides an overview of the *HBWW* initiative's strategies and anticipated outcomes.

HBWW was directed by a steering committee that provided organizational and financial oversight and an Executive Program Board that was responsible for program design, implementation, and data-related issues. A Kentucky site council composed of perinatal experts in Kentucky and representatives from the participating hospitals and departments of health was established to advise on all aspects of planning and implementation. Several work groups were also convened that focused on the specific program areas of interventions, communications, and evaluation (Appendix B).

Kentucky was selected for the *HBWW* initiative because it met a priori operational criteria including an electronic birth certificate system that conforms to the 2003 revision of the US Standard Birth Certificate, concerned clinical and public health leaders willing to take action to decrease preterm

Table 1 *HBWW* intermediate objectives

Community	Objective
Patients	Integrate clinical and support services. Provide consistent messaging. Create positive changes in knowledge, attitudes, and behavior regarding modifiable risks for preterm birth to promote healthy behaviors.
Perinatal providers	Bring the latest research to everyday practice with a focus on late preterm births. Create positive changes in practice regarding adherence to professional guidelines for preterm birth prevention and the elimination of elective deliveries before 39 weeks.
Public	Increase awareness of the risks, health consequences, and the importance of preventing preterm birth.

birth, and three program design criteria: (1) a preterm birth rate of 15%, above the 2006 national average of 12.8% [1], (2) a high and rapidly rising late preterm birth rate [18] that had increased more than 20% during the preceding decade, and (3) a relatively homogenous population with high rates of modifiable risk factors for preterm birth (smoking, substance use [19], and folic acid deficiency [20]).

Three IS, each consisting of a hospital-local health department dyad, were chosen to implement a bundle of preterm birth prevention interventions over a 3-year period. A *site* included a targeted hospital with its catchment area as well as the associated local health department. The IS and CS hospital groups each consisted of an academic referral center and two community hospitals that demonstrated high and increasing rates of preterm birth. IS were selected on the basis of their perinatal leadership, delivery volume (approximately 1000–2000 births annually), characteristics of women served (e.g., modifiable risk factors for preterm birth), geographic diversity, and willingness to participate. Each CS hospital was selected to be comparable to an IS hospital in delivery statistics and demographics, and to minimize contamination, each also had to be geographically distant (at least 70 miles) from any IS and located in a different county and health district. *HBWW* site locations are shown in Figure 1. The six paired IS and CS hospitals participating in *HBWW* were the following:

- Trover Health System (IS) in Madisonville paired with Western Baptist Hospital (CS) in Paducah *(Western KY)*;
- University of Kentucky Hospital (IS) in Lexington paired with Norton Hospital Downtown (CS) in Louisville *(Central KY)*;
- King's Daughters Medical Center (IS) in Ashland paired with Lake Cumberland Regional Hospital (CS) in Somerset *(Eastern KY)* (Table 2).

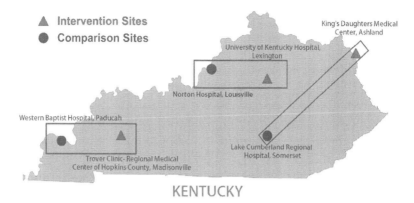

Figure 1 Healthy babies are worth the wait: hospital sites.

Table 2 Description of intervention sites (IS) and comparison site (CS) hospitals

Intervention hospitals		Local health departments	Comparison site hospitals	
Name	Description		Name	Description
King's Daughters Medical Center, Ashland, KY	Private Community Hospital 1315 births[a]	Ashland— Boyd County Health Department	Lake Cumberland Regional Hospital, Somerset, KY	Community Hospital 1482 births[a]
Trover Clinic- Regional Medical Center of Hopkins County, Madison- ville, KY	Community Hospital 980 births[a]	Hopkins County Health Department	Western Baptist Hospital, Paducah, KY	Community Hospital 1000 births[a]
University of Kentucky Hospital, Lexington, KY	Academic Center 1731 births[a]	Fayette County Health Department	Norton Hospital (Down-town Hospital), Louisville, KY	Academic Center 1414 births[a]

[a]Singleton inborn births occurring in July 2005–June 2006.

The *HBWW* initiative was composed of multiple evidence-based interventions chosen by an expert panel familiar with state and local service delivery systems. Several of these interventions, particularly those related to patient education and consumer awareness, were developed specifically for the *HBWW* program. Interventions primarily targeted perinatal providers, pregnant women, and women of child-bearing age and included: delivery of a communications campaign creating educational messaging for the public, provision of continuing education for providers, implementation of patient safety strategies, and enhancement of linkages of families to services in the community that provide culturally sensitive, health literacy appropriate information. The interventions were delivered through a system of collaboration among local and state-level clinical and public health partners. Each IS tailored and implemented preterm birth-related interventions involving patient education, provider education, consumer awareness, service integration, and clinical and public health interventions.

Interventions for *pregnant patients* sought to ensure that they had the information and support needed for healthy pregnancies. A signature *HBWW* brochure on the importance of the last month of pregnancy and a fetal brain development card ("brain card") were created with the help of community-based focus groups and were used at prenatal visits and outreach activities (Appendix C). IS hospitals and local health departments also collaborated to enhance existing case management, screening, and referral services for pregnant women to promote smoking cessation, treatment of other substance abuse, dental health, daily consumption of folic acid, stress reduction, and other healthy behaviors.

Provider interventions sought to ensure that each clinician had the latest information on preterm birth, late preterm birth, and implementation of evidence-based practices. In 2006, the Clinics in Perinatology published a series of articles in a monograph titled, *Late Preterm Pregnancy and the Newborn* [21] that described the epidemiology, physiology, and complications associated with late preterm infants. In particular, an article by Fuchs and Wapner identified the practice of elective cesarean births and inductions as contributing factors to the rise in late preterm births. This information guided the development of interventions for the *HBWW* initiative and was reinforced by the August 2009 ACOG Practice Bulletin No. 107 [12] that reaffirmed 39 weeks as the minimum gestational age for elective deliveries. The interventions consisted of Grand Rounds lectures and other continuing education presentations, notification of the latest research articles and professional guidelines, use of progesterone to prevent recurrent preterm birth in eligible patients [22], and quality improvement/patient safety efforts aimed at reducing elective inductions and cesarean births during the early term and preterm period.

The *HBWW* sites, in collaboration with the March of Dimes Kentucky Chapter and the Kentucky Department for Public Health also promoted *public* awareness of and engagement in preterm birth prevention through a media campaign (involving local newspaper, radio, and television), community outreach at health fairs and other public events, and educational presentations using a signature *HBWW* prematurity toolkit. Additionally, a Web site on preterm birth for consumers and providers (www.prematurityprevention.org) was developed and launched.

The interventions summarized below were bundled in varying combinations to meet the specific needs at each IS:

- Community awareness and education identifying the problem and consequences of preterm birth and late preterm birth, using community-based outreach, and promoting culturally sensitive, health literacy appropriate information within the context of prenatal care.
- Professional continuing education through Grand Rounds presentations, conferences, webinars, weekly relevant peer-reviewed publications, and other venues for a broad range of providers who serve pregnant women.
- Public health interventions augmenting existing services for home visitation, smoking cessation, psychosocial screening and referral, brief interventions for substance use, and in identified sites, colocation of clinical and public health staff/services.
- Clinical interventions principally in the prenatal period, complemented by preconception and interconception care:
  - Comprehensive prenatal care—per standard clinical guidelines (American Congress of Obstetricians and Gynecologists, American Academy of Pediatrics, Association of Women's Health, Obstetric and Neonatal Nurses),
  - Patient safety and quality improvement activities, including initiatives to eliminate elective deliveries before 39 weeks gestation,
  - Substance use prevention/cessation (tobacco, alcohol, other drugs),
  - Progesterone (17P) for prevention of recurrent preterm birth in eligible women [22],
  - Daily appropriate folic acid consumption [20],
  - Infection/inflammation prevention and management [3],
  - Promotion of oral health, such as preventing and managing periodontal disease,
  - Support for an existing evidence-based group prenatal care model (CenteringPregnancy®) [23–25] (Web site: www.centeringhealthcare.org).

Although findings on the relationship between oral health and preterm birth are not consistent [26,27], published medical literature at the time [28–30] supported the inclusion of oral health promotion as an intervention in this initiative.

At the conclusion of the *HBWW* pilot in Kentucky, five core program components were identified that succinctly characterized the essential activities of the initiative (Table 3). Figure 2 graphically demonstrates how the core program components interrelate to achieve the goal of preterm birth reduction in the *HBWW* initiative.

EVALUATION DESIGN

The *HBWW* evaluation included process and outcome components, and both a formative focus, in which the goal was program improvement, and a summative focus to determine whether *HBWW* achieved the intended outcomes. It was implemented through a collaboration between an external program evaluator, the lead initiative partners, and the evaluation work group which included Executive Program Board members, other partners, and advisors. The evaluation employed a "real world" mixed ecologic design of *HBWW* that combines basic features of the multiple-group study and the time-trend study to assess change over time within intervention and comparison groups, as well as differences across groups [16,17,31]. The group, consisting of the aggregated IS versus the aggregated CS—not the individual pregnant patient, provider, birth, or site—is the principal unit of analysis in the evaluation adhering to the ecologic design principle.

To assess program implementation, the team leveraged existing state and local service delivery data systems and developed additional tracking logs. To evaluate knowledge, attitudinal, and behavioral (KAB) outcomes relevant to *HBWW*'s theory of change, the team employed anonymous convenience sample surveys of consumers (pregnant women) and perinatal providers in the six *HBWW* sites. Interviews with local hospital and health department leaders provided a source of qualitative data on relevant environment and policy changes. To assess perinatal outcomes, birth certificate data from before, during, and after the *HBWW* initiative were analyzed.

Program Implementation

During the initiative, compliance and exposure to the interventions were evaluated to complement analysis of the perinatal outcomes. Tools used and data collected included tracking logs to document consumer and professional

Table 3 *HBWW* core components

Component	Description
1. Partnerships and collaborations	Significant success can be achieved through partnerships and collaborations. While these relationships may take time to develop, they can enhance the coordination, effectiveness, and efficiency of patient services and connect pregnant women to comprehensive medical and public health services.
2. Provider initiatives	Provider education on preterm birth prevention is essential. It should emphasize current research on the consequences of late preterm birth, and offer resources for providers to implement interventions based on this information in order to speed research to practice. Providers include all hospital and public health professionals who work with prenatal patients.
3. Patient support	Pregnant women and women of childbearing age need to be informed about the problem of preterm birth and the associated risk factors. Support services targeting risk reduction should be provided and resources for building protective factors made available. Patient education should be culturally and health literacy appropriate and delivered by clinical and public health providers. Educational messages can be provided through multiple media channels and at multiple points of contact.
4. Public (community) engagement	Preterm birth affects pregnant women and their infants as well as their family, friends, and the broader community. Nationally, there are significant economic costs related to both the short- and long-term consequences of preterm birth. Raising awareness about the multiple ramifications of preterm birth through a variety of media and communication strategies increases the prospect of securing resources to address the problem, facilitates the dissemination of important prevention messages to pregnant women, and heightens awareness about the role that family, friends, and the community have in supporting pregnant women in their efforts to have a healthy pregnancy and healthy infant.
5. Progress measurement	Documenting the impact of the *HBWW* initiative is critical and requires careful consideration of the evaluation methods employed. Measures to assess progress should be predetermined and mechanisms to collect the data should be in place before implementing the program. Each collaborating member organization participating in the partnership should be responsible for providing data on their activities. It is important to track progress throughout the course of the program and celebrate successes along the way.

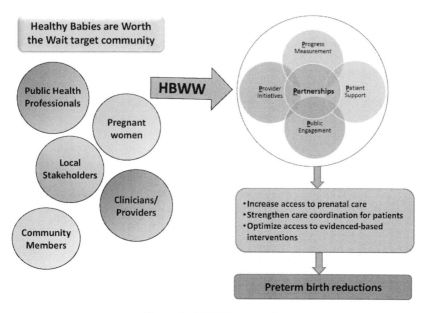

Figure 2 *HBWW* framework.

compliance with the interventions; standard industry reports to characterize media coverage by market; administrative data to count visits to the *HBWW* Web site; and counts of trainings, educational conferences, Grand Rounds lectures, and program materials distributed (Appendix D). Investigators also monitored the frequency of prenatal service home visits, referrals to the Kentucky Tobacco Quit Line, and use of the *HBWW* toolkit (a resource used to facilitate community education about preterm birth).

Consumer and Provider Surveys

Surveys of consumers (pregnant women) and providers were conducted at the initiative's baseline period (January–May 2007) and at follow-up (July–November 2009) in the six IS and CS hospital catchment areas, using anonymous voluntary convenience samples at each time point. The consumer and provider surveys were both designed to collect formative data to guide the development and evolution of *HBWW* (Appendix E). The consumer survey was also used to test the hypotheses that from baseline to follow-up, pregnant women using prenatal services in the IS catchment areas would show more favorable improvements in preterm birth-related KAB, and that at follow-up they would report greater exposure to *HBWW* than their counterparts in the CS catchment areas. In addition, the survey was designed to provide some data on selected preterm birth-related messages that these patients received from their providers. The provider survey was used to test the hypotheses that from baseline to follow-up, IS providers would show more favorable improvements in preterm birth-related knowledge and attitudes and that at follow-up they would report greater exposure to *HBWW* than their counterparts in the CS.

The consumer survey employed closed- and open-ended items, including questions drawn from the Centers for Disease Control and Prevention's (CDC) Pregnancy Risk Assessment Monitoring System (PRAMS), CDC's Behavioral Risk Factor Surveillance System (BRFSS), and March of Dimes' Gallup Surveys, and some questions developed specifically for the *HBWW* initiative. The survey was made available in English and Spanish. Participants were 1066 pregnant women at baseline (765 from the IS and 301 from the CS) and 1122 pregnant women at follow-up (598 from the IS and 524 from the CS). The participants were recruited by trained administrative and nursing staff in the waiting areas of prenatal care facilities in the six *HBWW* hospital catchment areas.

The provider survey was administered to physicians, nurses, and advance practice nurses and employed closed- and open-ended items, including some questions drawn from the 2005 Society of Maternal and Fetal Medicine Progesterone Survey [32] and the New York Chapter (District II) of the American Congress of Obstetricians and Gynecologists (ACOG) 2006 Preterm Survey [33]. It also contained questions developed specifically for the *HBWW* initiative. Participants were 135 perinatal providers at baseline (101 from the IS and 34 from the CS) and 192 perinatal providers at follow-up (97 from the IS and 95 from the CS). Providers who worked with pregnant women were approached about the survey by administrative staff at each IS and CS, under the direction of the *HBWW* site leads.

For the summative evaluation, univariate analyses were run on all closed-ended and numerical response items in the consumer surveys, by intervention versus comparison group. For these variables, analyses included frequency distributions and percent change from baseline to follow-up. Chi-square tests or *t*-tests (as appropriate) were also performed for key *HBWW* exposure and KAB variables to assess differences between the intervention and comparison groups at each time point. For the KAB variables, logistic regression analyses were performed in which key maternal covariates were controlled for including race (White, Black, Other), education (<high school, high school graduate/GED, >high school), previous birth (yes, no), and trimester of pregnancy. Effect sizes at follow-up for intervention versus comparison group were also calculated using the parameter estimates from the logistic regression models. Given the characteristics of the study design, it was not appropriate to assess changes over time in the intervention or comparison groups for statistical significance, nor was it appropriate to compare intervention and comparison group trajectories over time using tests of statistical significance.

For the provider survey, univariate analyses were run on all closed-ended and numerical response items in the all-provider data set as well as in the physician-only subset, by intervention versus comparison group. For the all-provider data set, chi-square tests or *t*-tests (as appropriate) were performed on key *HBWW* exposure and knowledge and attitudinal outcome variables to assess differences between the intervention and comparison groups at each time point. The physician-only sample size was too small to appropriately conduct and interpret such analyses. As in the case of the consumer survey, given the characteristics of the study design, it was not appropriate to assess changes over time in intervention or comparison groups for statistical significance, nor was it appropriate to compare intervention and comparison group trajectories using tests of statistical significance.

Analyses were performed using SPSS® version 16.0 or SAS® version 9.1.3, depending on the specific analysis. Missing data were excluded from all analyses. Statistical significance was set at $p < 0.05$. Institutional Review Board (IRB) exemption was granted by the Kentucky Cabinet for Health and Family Services and each site obtained local IRB approval or exemption to conduct surveys.

Annual Perinatal Systems Survey and Environment and Policy Interview

An Annual Perinatal Systems Survey (APSS) was conducted from 2007 to 2009 with the administrators from the three IS and three CS hospitals. These surveys provided a key source of qualitative data on relevant contextual changes that might impact implementation of *HBWW* as well as perinatal outcomes in the sites' catchment areas. Using a combination of closed- and open-ended items, the APSS collected largely qualitative information about the services available for pregnant women in each hospital's catchment area, continuing education offerings for obstetrical providers, hospital policies related to preterm birth, and the hospital obstetric patient safety/quality improvement program, as well as some basic statistics on annual prenatal care patients and visits (Appendix F).

A semistructured Environment and Policy Interview (EPI) protocol was also implemented annually with a representative from each of the six hospital sites and (starting in 2008) a representative from each of the three IS local health departments to collect further qualitative data on changes with respect to services for pregnant women, hospital policies, and other local, state, or national policies, services, economic factors, and social circumstances that might influence preterm-related services or preterm birth rates. In the final EPI conducted in 2009, a question about involvement in and exposure to *HBWW* activities was added to the CS protocol (Appendix G).

Issues of staff turnover, varied interpretation of the APSS questions between sites, and varied recall of information from 1 year to the next supported the use of the EPI as an important complement to the APSS. The data collected were organized by site and by topic and analyzed for key themes and trends.

Perinatal Outcomes

To permit assessment of perinatal outcomes the Kentucky Department for Public Health Office of Vital Statistics made available natality files for the period July 2005 through June 2010. These data files provided detailed birth information based on the 2003 revision of US Standard Birth Certificate. Analysis was limited to singleton, inborn (births in which the mother was not transferred into the hospital as indicated by a checkbox on the birth certificate) live births. Gestational age in completed weeks of pregnancy was assigned to each birth record in accordance with methodology established by the National Center for Health Statistics, based on the calculation of the interval between the first day of the mother's last menstrual period and the date of birth, and described in detail elsewhere [1].

The primary outcome measures examined were rates of preterm birth and late preterm birth with measurements taken pre- and post-*HBWW* implementation. Analyses of these measures and their changes over time compare the three aggregated IS to the three aggregated CS and to the aggregated remaining hospitals in Kentucky. This mixed ecologic design [17] is limited in that it does not allow for the isolation of any specific intervention within the bundle for further analysis, and it does not adequately account for the possibility of external causation over time; but as a public health demonstration project, the project objective to reduce preterm birth in Kentucky superseded research concerns.

The pre-implementation baseline period (July 2005–June 2006) was identified as the 12 months prior to the July 2006 start of planning and training activities in the IS hospitals; and the post-implementation period was defined as July 2009 through June 2010 (Figure 3). Significant changes between the

July 2005–	July 2006–	July 2007–	July 2009–
June 2006	June 2007	June 2009	June 2010
Pre-implementation baseline period: perinatal outcomes	*Planning, training, and baseline survey data collection*	*Implementation of principal interventions under March of Dimes and Johnson & Johnson Pediatric Institute, LLC funding*	*Post-implementation period: follow-up survey data collection & perinatal outcome data analysis*

Figure 3 Project timeline.

pre- and post-implementation periods were determined at a p-value of 0.05 using a chi-square test.

Logistic regression was used to estimate the reduction in odds of preterm birth between pre- and post-implementation. To determine differences among time effects across hospital groups, the model included both individual variables for the pre- and post-implementation periods, and an interaction term. Analyses controlled for maternal age (<20 years, 20–29 years, 30 years and older), race (White, Black, Other), and history of previous preterm birth (yes, no). Data analyses were conducted with SAS® software version 9.1.3 (SAS Institute Inc., Cary, NC) and SPSS® version 15.0 (SPSS, Inc., Chicago, IL).

CHAPTER 4

Healthy Babies are Worth the Wait® Planning and Implementation Activities

Three focus groups were conducted with pregnant women from several communities in Kentucky prior to the implementation of *Healthy Babies are Worth the Wait*® *(HBWW)*. The results indicated that preterm birth was not a major concern for the participants, and that their definition of "term pregnancy" ranged from 30 to 40 weeks. Information obtained from these focus groups was used to develop consumer education resources for pregnant women. A brochure entitled *Every Week Counts* was created to inform mothers about the importance of continuing pregnancy to at least 39 weeks in the absence of medical indications; and an education tool, *Brain Growth Matters* ("brain card"), was developed for providers to counsel their pregnant patients about important brain development occurring in the last weeks of pregnancy.

Patient education about prevention and the potential complications of late preterm birth, avoiding elective delivery prior to 39 weeks, and other *HBWW* prevention messages were incorporated into routine prenatal education and promoted at health fairs and community events. A *HBWW* Web site provided a conduit for pregnant women to obtain information about *HBWW* and late preterm birth and afforded access to an interactive pregnancy diary for mothers to record their pregnancy progress and develop a birth plan. Computer access to the Web site was available to women during their prenatal care visits at the intervention site hospitals. Examples of the *HBWW* consumer education products and marketing materials are provided in Appendix C.

Community outreach activities and local media coverage provided important opportunities to raise awareness about preterm birth in Kentucky communities. A *HBWW* toolkit for "Community partners to take action to prevent preterm birth" was developed collaboratively by the Kentucky Department for Public Health (KYDPH) and March of Dimes for use in community workshops. The toolkit provides information about preterm birth in Kentucky, identifies community activities to promote prematurity prevention, and includes a media guide to facilitate dissemination of preterm birth prevention messages. Presentations using the toolkit were conducted at multiple venues in the intervention sites (IS) including day care centers, women's groups, libraries, and by prenatal case managers and public health nurses during home visits and clinic appointments. *HBWW* materials that conveyed the initiative's key messages (Table 4) were distributed at several cultural festivals and health fairs, and at other fundraising, entertainment, and sports events.

Broad exposure of the *HBWW* initiative nationally was facilitated through several important policy forums, including the 2008 Surgeon General's Conference on the Prevention of Preterm Birth, the 2009 March of Dimes Symposium on Quality Improvement to Prevent Prematurity, and the 2009 Eunice Kennedy Shriver National Institute of Child Health and Human Development, National Child and Maternal Health Education Program Coordination Committee meeting. In 2009, a special session on *HBWW* was held at the American Public Health Association Annual Meeting, and presentations were made at other national forums including the Association of Maternal and Child Health Programs, CityMatCH, and the Maternal and Child Health Epidemiology Program. The national audience for these presentations resulted in broad dissemination of the initiative's focus and generated significant momentum to reduce preterm births on a national level.

Extensive provider trainings were held at each of the three IS, and continuing education opportunities were offered for a diverse audience of perinatal providers, public health staff, dentists, risk managers, and hospital administrators. Provider educational resource centers were established at each intervention site that facilitated easy access to clinical guidelines, protocols, and monthly updates of relevant journal articles on preterm birth. *HBWW* sponsored Grand Rounds lectures, webinars, and presentations at state conferences (Kentucky Perinatal Association (KPA), Kentucky Association of Women's Health, Obstetric and Neonatal Nurses, Neonatal Nursing Conferences, and Prematurity Summits), reaching more than 3500 perinatal providers. *HBWW* also provided funding support to the KPA's Web-based educational program, Health Professional

Table 4 *HBWW* key messages

- See your doctor *before* and *during* pregnancy
- Avoid scheduled delivery before 39 weeks of pregnancy unless medical problems make it necessary
- Get help to stop smoking and avoid secondhand smoke
- Do not use alcohol or drugs
- Take folic acid *every* day
- Brush, floss, and visit the dentist
- Important brain growth occurs in the baby during the last weeks of pregnancy

Education on Prematurity, which offered free educational credits to perinatal providers (www.kentuckyperinatal.com/hpep.html). *HBWW* Newsletters were published biannually and disseminated at the intervention sites to keep providers informed about the initiatives activities and accomplishments (Appendix H).

In accordance with *HBWW's* ecologic design, intervention content and emphasis varied across the three IS. Cultural changes in clinical practice, catalyzed by exposure to the continuing education resources, were evidenced at each IS hospital by the adoption of quality improvement (QI) procedures to address elective deliveries. QI practices evolved during the third year of the initiative following extensive education of providers and hospital administrators during the second year. The IS nursing staff working with *HBWW* successfully engaged hospital administrators and obtained buy-in from providers to implement QI protocols to prevent elective deliveries before 39 weeks. Wide variations in provider practices at the IS were successfully modified following the development of practice policies and QI procedures. Surveillance and reports of physicians' clinical practices provided useful feedback to the doctors and identified when corrective actions were required.

Service enhancements to existing programs that supported improved perinatal outcomes were also facilitated by the *HBWW* initiative. Examples of these service enhancements include the *HBWW* funding to support the implementation of a quality assurance "site approval process" for CenteringPregnancy® [23–25], an empirically validated group prenatal care model, at two of the IS. Expansion of social work services to address the burgeoning abuse of prescription painkillers was supported at one IS hospital through a grant from the Kentucky March of Dimes chapter.

Successful collaborations between the IS hospitals and public health services occurred during the *HBWW* initiative. Information on the smoking cessation resources of the KYDPH was provided to the hospitals. This resulted in enhanced counseling by providers and facilitated referrals to the Kentucky Quit Line. At one site, the KYDPH Health Access Nurturing Development Services (HANDS) perinatal home visiting program was expanded to include families that were experiencing a repeat pregnancy versus only serving first-time parents. Psychosocial screening was built into the prenatal record at one IS hospital, and a fax referral form was created to facilitate referrals from the IS hospital to the local health departments to coordinate patient services. Every IS enhanced or developed new methods of communication and referrals between the public health and hospital teams.

CHAPTER 5

Evaluation Results

CONSUMER AND PROVIDER SURVEYS

Consumer Survey

As presented in Table 5, a total of 1066 baseline consumer surveys and 1122 follow-up consumer surveys were included in the summative analyses. The intervention and comparison group survey respondents showed some significant differences at baseline and/or follow-up with respect to various demographic and pregnancy/birth history factors (such as race, ethnicity, education, history of preterm birth, and trimester in which the survey was completed); as was noted above, selected factors were controlled for in the logistic regression analyses of the key knowledge, attitudes, and behaviors (KAB) outcome variables.

With respect to *Healthy Babies are Worth the Wait*® *(HBWW)* exposure, at follow-up the intervention group was significantly more likely than the comparison group to report having heard of *HBWW* (34.5% vs 7.9%, $p < 0.01$), seen the "brain card" (22.9% vs 12.7%, $p < 0.01$), and received at least one of six items with *HBWW* or the *HBWW* Web site on it (47.2% vs 8.4%, $p < 0.01$). However, reported usage of the *HBWW* Web site was very low across both groups—less than 3% among the intervention group and less than 1% among the comparison group[1]. It is unclear whether this low usage was due to insufficient Web site marketing, competition with higher profile Web sites targeting pregnant women, limited Internet access, or other factors.

For the summative evaluation, seven key consumer KAB outcomes were examined:

1. Knowledge of length of a term pregnancy.
2. Knowledge of whether a cesarean birth is usually safer than a vaginal birth.
3. Knowledge of whether women tend to have more problems later in life if they have a vaginal (vs cesarean) birth.
4. Knowledge of whether it is a good idea to schedule delivery at 35–36 weeks for convenience.

---

[1] The comparison sites (CS) sample distribution did not permit appropriate statistical comparison of the intervention sites (IS) and CS groups.

Table 5 Number of consumer surveys

Baseline (N=1066)						Follow-up (N=1122)					
IS (N=765)			CS (N=301)			IS (N=598)			CS (N=524)		
KD	TC	UK	LC	NH	WB	KD	TC	UK	LC	NH	WB
289	337	139	57	93	151	169	248	181	216	150	158

Key: KD = King's Daughters; LC = Lake Cumberland; NH = Norton Hospital; TC = Trover Clinic; UK = University of Kentucky; WB = Western Baptist.

5. Attitude toward how serious a problem preterm birth is in the community.
6. Any smoking during current pregnancy.
7. Any drinking during current pregnancy.

The data suggest (Table 6) that over time the intervention group fared better than the comparison group with respect to knowledge and attitudes (items 1–5, above).

In the unadjusted analyses, for knowledge of term pregnancy length, IS respondents had a significantly lower percentage correct than the CS respondents at baseline (86.9% vs 94.0%, $p < 0.01$), but the groups did not differ significantly at follow-up (89.6% vs 89.4%, $p = 0.89$) (Figure 4a).

For knowledge of whether it is a good idea to schedule delivery for convenience at 35–36 weeks, the pattern was similar: the IS had a significantly lower percent correct than the CS at baseline (61.4% vs 72.5%, $p < 0.01$), but at follow-up the groups did not differ significantly (67.1% vs 66.9%, $p = 0.99$) (Figure 5a).

For knowledge of whether women have more problems later if they have a vaginal (vs cesarean) birth, the groups did not differ significantly at baseline, but at follow-up the IS had a significantly higher percentage correct than the CS (69.7% vs 63.1%, $p = 0.02$) (Figure 6a).

For the remaining key knowledge variable (knowledge of whether a cesarean birth is usually safer than a vaginal birth), IS and CS respondents' percentages of correct responses did not differ significantly at baseline ($p = 0.14$) or at follow-up ($p = 0.08$) (Figure 7a).

For all four knowledge variables, the intervention group showed a positive change in percent correct over time, while the comparison group showed a negative change.

In the unadjusted analyses of the attitudinal variable (how serious a problem preterm birth is in the community), response distribution (for considering it to be a "serious" problem) was not significantly different between the IS and CS at baseline (37.9% vs 35.7%, $p = 0.50$), but at follow-up a significantly larger percentage of IS than CS respondents reported it was

Table 6 Key consumer KAB outcome variables: frequencies and change over time[a]

KAB variable (referent value)[a]	Unadjusted or adjusted[b]	Baseline (N = 1066)			Follow-up (N = 1122)			Change over time	
		IS (N = 765)	CS (N = 301)	p	IS (N = 598)	CS (N = 524)	p	IS	CS
1. Length of term pregnancy (correct)	Unadjusted	86.9%	94.0%	<0.01*	89.6%	89.4%	0.89	+3.1%	−4.9%
	Adjusted	88.0%	93.2%	0.62	89.9%	89.1%	0.19	+2.2%	−4.4%
2. Cesarean usually safer than a vaginal delivery (T/F) (correct)	Unadjusted	68.9%	73.5%	0.14	73.6%	68.9%	0.08	+6.8%	−6.3%
	Adjusted	70.5%	73.4%	0.75	74.4%	69.2%	0.01*	+5.5%	−5.7%
3. Women have more problems later if vaginal delivery (T/F) (correct)	Unadjusted	66.3%	67.5%	0.72	69.7%	63.1%	0.02*	+5.1%	−6.5%
	Adjusted	68.0%	67.7%	0.43	70.4%	64.6%	0.01*	+3.5%	−4.6%
4. Good idea to schedule delivery at 35–36 weeks for convenience (T/F) (correct)	Unadjusted	61.4%	72.5%	<0.01*	67.1%	66.9%	0.99	+9.3%	−7.7%
	Adjusted	63.6%	72.2%	0.28	67.6%	67.5%	0.30	+6.3%	−6.5%

Continued

Table 6 Key consumer KAB outcome variables: frequencies and change over time[a]—cont'd

KAB variable (referent value)[a]	Unadjusted or adjusted[b]	Baseline (N=1066)			Follow-up (N=1122)			Change over time	
		IS (N=765)	CS (N=301)	p	IS (N=598)	CS (N=524)	p	IS	CS
5. How serious a problem preterm birth is in community (*serious*)	Unadjusted	37.9%	35.7%	0.50	49.3%	42.5%	**0.02***	+30.1%	+19.0%
	Adjusted	37.9%	35.4%	0.82	48.9%	43.1%	0.08	+29.0%	+21.8%
6. Any smoking in current pregnancy (*none*)	Unadjusted	75.2%	78.9%	0.26	79.2%	79.9%	0.79	+5.3%	+1.3%
	Adjusted	74.8%	80.3%	0.51	79.3%	80.1%	0.97	+6.0%	−0.2%
7. Any drinking in current pregnancy (*none*)	Unadjusted	98.1%	99.6%	0.12	99.1%	98.4%	0.30	+1.0%	−1.2%
	Adjusted[c]	—	—	—	—	—	—	—	—

*Statistically significant at p < 0.05.

[a]For the analyses, responses were dichotomized to two values (correct or incorrect; serious or not serious/not sure; none or any). Missing values were excluded from analysis.

[b]The following covariates were controlled for in the adjusted analyses: race/ethnicity, education, history of previous birth, and trimester of pregnancy in which survey was completed.

[c]Because reported abstinence from alcohol during pregnancy was virtually at ceiling among both IS and CS respondents, at both time points, adjusted frequencies were not calculated for this variable.

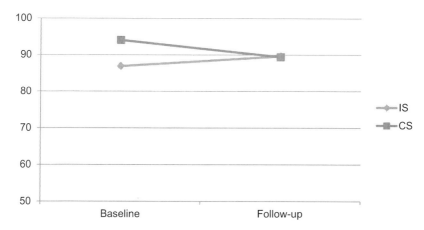

Figure 4a Respondents who correctly answered question on "Length of term pregnancy": Unadjusted percentages. Difference between IS and CS was statistically significant at baseline ($p < 0.01$). Difference between IS and CS was not statistically significant at follow-up ($p = 0.89$).

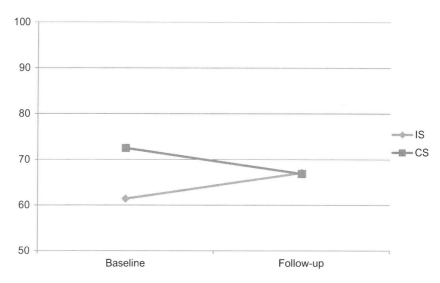

Figure 5a Respondents who correctly answered question on "Good idea to schedule delivery for convenience at 35–36 weeks for convenience": Unadjusted percentages. Difference between IS and CS was statistically significant at baseline ($p < 0.01$). Difference between IS and CS was not statistically significant at follow-up ($p = 0.99$).

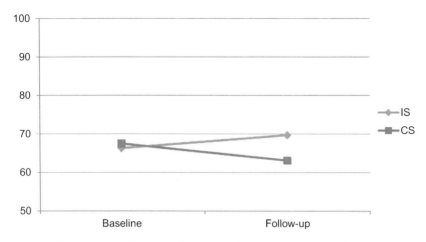

Figure 6a Respondents who correctly answered question on "Women have more problems later if vaginal delivery": Unadjusted percentages. Difference between IS and CS was not statistically significant at baseline (p = 0.72). Difference between IS and CS was statistically significant at follow-up (p = 0.02).

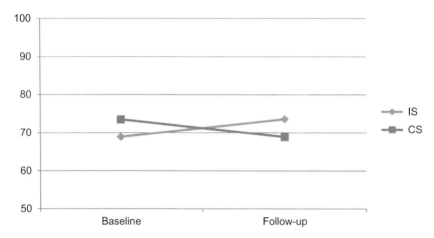

Figure 7a Respondents who correctly answered question on "Cesarean birth is usually safer than a vaginal birth": Unadjusted percentages. Difference between IS and CS was not statistically significant at baseline (p = 0.14) or follow-up (p = 0.08).

"serious" (49.3% vs 42.5%, p = 0.02). Both groups showed a positive change over time in the percentage that considered preterm birth to be a "serious" problem, but the intervention group's change (30.1%) was much larger than the comparison group's change (19.0%) (Figure 8a).

When key covariates were controlled for, the data also showed some better knowledge outcomes among the IS versus CS respondents.

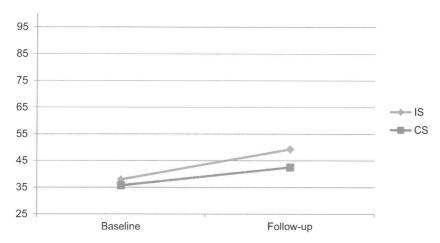

Figure 8a Participants who reported preterm births are a "serious" problem in their community: Unadjusted percentages. Difference between IS and CS was not statistically significant at baseline (p = 0.50). Difference between IS and CS was statistically significant at follow-up (p = 0.02).

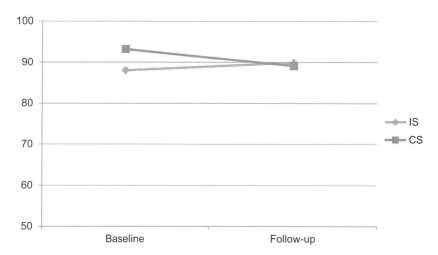

Figure 4b Respondents who correctly answered question on "Length of term pregnancy": Adjusted percentages. Difference between IS and CS was not statistically significant at baseline (p = 0.62) or follow-up (p = 0.19).

Specifically, at baseline no statistically significant differences between intervention and comparison groups emerged for any of the five items (Figures 4b–8b). At follow-up, a significantly greater percentage of intervention than comparison respondents knew that women do *not* tend to have more problems later with a vaginal (vs cesarean) birth

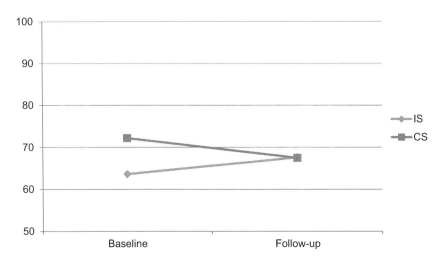

Figure 5b Respondents who correctly answered question on "Good idea to schedule delivery at 35–36 weeks for convenience": Adjusted percentages. Difference between IS and CS was not statistically significant at baseline (p = 0.28) or follow-up (p = 0.30).

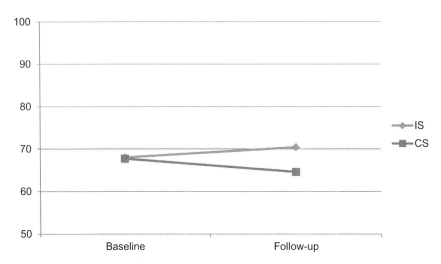

Figure 6b Respondents who correctly answered question on "Women have more problems later if vaginal delivery": Adjusted percentages. Difference between IS and CS was not statistically significant at baseline (p = 0.43). Difference between IS and CS was statistically significant at follow-up (p = 0.01).

(70.4% vs 64.6%, p = 0.01) (Figure 6b), and knew that a cesarean is *not* generally safer than a vaginal birth (74.4% vs 69.2%, p = 0.01) (Figure 7b). For other knowledge outcomes and for the attitudinal outcomes, IS and CS did not differ significantly at baseline or at follow-up.

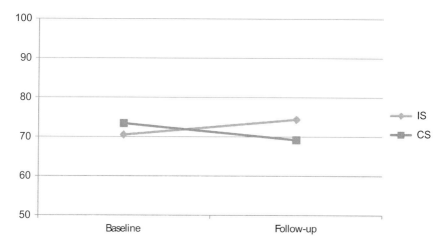

Figure 7b Respondents who correctly answered question on "Cesarean usually safer than a vaginal delivery": Adjusted percentages. Difference between IS and CS was not statistically significant at baseline (p = 0.75). Difference between IS and CS was statistically significant at follow-up (p = 0.01).

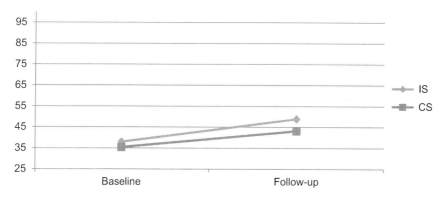

Figure 8b Participants who reported preterm births are a "serious" problem in their community: Adjusted percentages. Difference between IS and CS was not statistically significant at baseline (p = 0.82) or follow-up (p = 0.08).

As in the case of the unadjusted frequency data, for the knowledge variables, percent correct increased from baseline to follow-up among the intervention group but decreased among the comparison group. For the attitudinal variable, both groups showed an increase in the desirable attitude (i.e., that the problem is "serious"), but the increase was greater among the intervention group than among the comparison group (Figure 8b).

It should be noted that the validity of the unadjusted consumer knowledge and attitudinal data is limited by the fact that covariates were not controlled for. The validity of the adjusted data is limited by the fact that missing values are relatively high (8–13% across time points and groups, because even a single missing covariate caused a case to be excluded), and sample sizes are correspondingly smaller, making it more difficult to achieve statistical significance. However, the positive findings in both sets of analyses lend credence to the interpretation that over time the intervention group performed better with respect to knowledge and attitudes than the comparison group. Data on *HBWW* exposure, which strongly indicate that exposure was greater among intervention than comparison group members (see above), provide some evidence that the observed differences were likely due—at least in part—to the initiative.

With respect to key consumer behaviors (i.e., any smoking during current pregnancy, any drinking during current pregnancy), there were no significant differences between intervention and comparison groups at either time point, in any of the analyses. The intervention group had (somewhat) more favorable change-over-time percentages than the comparison group, but these small differences may simply be artifacts of the anonymous convenience sampling methodology and not indicative of actual community-wide changes (Figures 9a and 9b).

It should also be noted that reported abstinence from alcohol use in current pregnancy was already nearly at ceiling at baseline for both groups (98.1% for the IS and 99.6% for the CS), limiting room for positive change. Additionally, the possible effects of social desirability bias on the self-reported data are unknown, although the fact that the *HBWW* surveys were anonymous likely helped to reduce bias.

For the key consumer KAB variables, effect sizes for intervention versus comparison group at follow-up were low (according to Cohen's *d* scale), ranging from −0.02 to 0.12. Given the challenges of changing consumer KAB community-wide, this was not unexpected.

Additional consumer data also shed light on some provider behaviors—specifically messages that providers gave their patients about preterm birth-related services and behaviors. In particular, although intervention group respondents had, on average, significantly fewer prenatal care visits than comparison group respondents at the time they completed the follow-up survey, a significantly greater proportion of intervention respondents reported that a provider had spoken to them about each of four key services (KY Quitline, HANDS home visitation program, dental services, domestic violence services; $p < 0.01$ for each service) during the current pregnancy. (Note that baseline data on discussion of services between consumers and

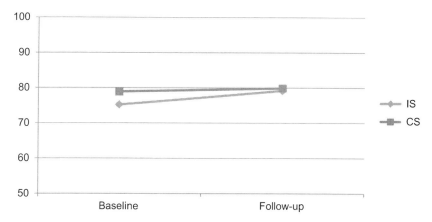

Figure 9a Participants who reported they do not smoke in current pregnancy: Unadjusted percentages. Difference between IS and CS was not statistically significant at baseline (p = 0.26) or follow-up (p = 0.79).

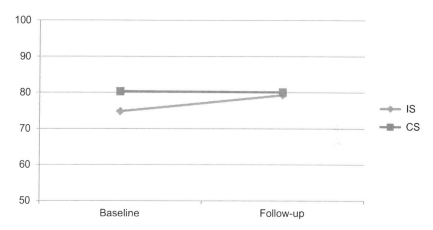

Figure 9b Participants who reported they do not smoke in current pregnancy: Adjusted percentages. Difference between IS and CS was not statistically significant at baseline (p = 0.51) or follow-up (p = 0.97).

providers were not collected.) In addition, although there were no significant differences between the intervention and comparison groups at baseline with respect to both whether providers had said not to smoke during the current pregnancy (p = 0.38), and whether providers had said not to drink alcohol during the current pregnancy (p = 0.20), at follow-up the intervention consumers were significantly more likely to report that both provider behaviors had taken place during the current pregnancy than were comparison consumers (p = 0.04 for discussion of not smoking, p < 0.01 for discussion of not drinking alcohol). This is consistent with intervention versus comparison group providers' reported exposure to *HBWW* and use of *HBWW* materials (see below). (As further context, it should be noted that at neither baseline nor follow-up did intervention and comparison groups differ significantly with respect to reports of any smoking or any alcohol use in the month before or anytime during the current pregnancy.)

Provider Survey

A total of 135 baseline provider surveys and 192 follow-up provider surveys were included in the summative analyses (Table 7). Overall, sample sizes for the intervention and comparison groups (and for physician respondents, in particular) were relatively small, and the distribution of physicians versus other providers varied greatly across the individual sites. The small sample sizes limited the range of analyses that could appropriately be performed.

With respect to *HBWW* exposure, among the all-provider data set (i.e., physicians, nurses, and others), the intervention group was significantly more likely than the comparison group to report having heard of *HBWW* (91.5% vs 44.1%, p < 0.01), used at least one of eight *HBWW* materials (69.9% vs 11.3%, p < 0.01), participated in *HBWW* continuing education activities (44.0% vs 9.5%, p < 0.01), and used the *HBWW* Web site (23.1% vs 6.7%, p < 0.01). Among the subset of providers who identified themselves as physicians, the exposure differences between intervention and comparison group are even more striking: among intervention physicians, 100% had heard of *HBWW*, while among comparison physicians only a third had heard of it. In addition, among intervention physicians, over half had used at least one of eight *HBWW* materials, over two-thirds had participated in *HBWW* continuing education, and over a tenth had ever used the *HBWW* Web site. No comparison physicians reported having ever used any of these resources. However, because the physician follow-up survey sample sizes are very small (19 in the intervention group, 9 in the comparison group), these frequencies should be interpreted with caution.

For the summative evaluation, five key provider knowledge and attitudinal outcome variables were examined:

1. Knowledge of gestational age for elective induction per ACOG guidelines.
2. Concern about increasing rates of late preterm birth.
3. Recommended mode of delivery for self/spouse if having a low-risk pregnancy.

Table 7 Number of provider surveys

| | Baseline (N=135) | | | | | | Follow-up (N=192) | | | | | |
| | IS (N=101) | | | CS (N=34) | | | IS (N=97) | | | CS (N=95) | | |
	KD	TC	UK	LC	NH	WB	KD	TC	UK	LC	NH	WB
All perinatal providers	16	32	53	5	4	25	44	21	32	16	25	54
Physicians	0	2	21	4	4	1	5	0	14	1	4	4
Other providers[a]	16	29	30	0	0	23	39	19	16	15	21	49
Profession missing	0	1	2	1	0	1	0	2	2	0	0	1

Key: KD = King's Daughters; LC = Lake Cumberland; NH = Norton Hospital; TC = Trover Clinic; UK = University of Kentucky; WB = Western Baptist.
[a]Includes nurses, nurse practitioners, nurse-midwives, and others.

4. At what gestational age would want self/spouse to deliver if having a low-risk pregnancy.

5. Whether would consider elective induction for self/spouse.

Three of the questions were personalized (items 3–5) in an effort to better elicit the attitudes of the respondents by asking what they would want or recommend for themselves or their spouse. The pattern of KAB findings in the all-provider data set is difficult to interpret. For three variables (items 2–4), the intervention and comparison groups did not differ significantly at either time point. For one variable (item 1), the intervention group had a significantly better outcome than the comparison group at both baseline and follow-up. For the remaining variable (item 5), the intervention group had a significantly worse profile than the comparison group at baseline but was not significantly different from the comparison group at follow-up. While this last pattern aligns with the expected findings, it is clearly the exception among the variables examined. For four of the five variables (items 2–5), the all-provider intervention group showed an improvement from baseline to follow-up. The all-provider comparison group showed an improvement for only two variables (items 1 and 4); however, for both, the comparison group's improvement far exceeded the intervention group's improvement. Overall, the varying provider data patterns are likely due—at least in part—to the combination of (1) the anonymous convenience sampling methodology, in which not all of the same providers were surveyed at each time point, participation by site and institution varied over time, and individual-level changes over time could not be tracked; and (2) the small sample sizes, which made the findings particularly prone to variation across time points and precluded controlling for covariates. In addition, despite the surveys being anonymous, social desirability bias may have impacted responses.

As was indicated above, the physician-only sample sizes were too small to appropriately compare intervention and comparison group outcomes, using tests of statistical significance, at either time point. Overall, the direction of change over time was similar among both the intervention and comparison groups. The only item for which there was a notable difference between groups was item 5 above (whether respondent would consider elective induction for self/spouse); for this variable there was an increase in the percent of the intervention group reporting that they would not consider an elective induction for self/spouse, and a decrease among the comparison group.

At least one factor outside the purview of *HBWW* may have had some impact on physician follow-up responses. The August 2009 ACOG Practice Bulletin [12] (which was issued as the follow-up provider survey was beginning administration) reaffirmed 39 weeks as the minimum gestational age for elective procedures. Both intervention and comparison group physicians showed increases from baseline to follow-up in knowledge of this guideline, as was indicated above.

Despite the equivocal findings of the provider survey, there is some stronger evidence from the consumer survey that intervention group perinatal providers exhibited some changes in behavior. In particular, as discussed above, there is evidence that intervention providers (in contrast to comparison group providers) increased provision of messages and information to patients about services and behaviors that could help to reduce preterm birth. This is consistent with intervention group providers' much greater reported exposure to *HBWW*.

Overall, the consumer and provider survey analyses strongly suggest that *HBWW* had a positive impact on consumer knowledge and on provider provision of messages and information to patients about services and behaviors that could help to reduce preterm birth, in the intervention site catchment areas. Findings concerning perinatal outcomes (see below) provide an additional assessment of the potential impact of *HBWW* on preterm birth rates in the IS.

ANNUAL PERINATAL SYSTEMS SURVEY AND ENVIRONMENTAL AND POLICY INTERVIEWS CROSS-SITE SUMMARY OF FINDINGS

The Annual Perinatal Systems Survey (APSS) and Environmental and Policy Interviews (EPIs) yielded important contextual information about the IS and CS catchment areas, including evolution of smoking policies and smoking cessation services; hospital policies and strategies concerning early elective deliveries; hospital policies concerning use of progesterone to prevent recurrent preterm birth; educational, health, and social services for pregnant women; other environmental factors relevant for *HBWW* implementation and perinatal outcomes; and CS exposure to *HBWW*.

Smoking Policies and Cessation Services

Largely as a result of legislative policy to enforce smoke-free ordinances, all of the IS and CS hospitals and their local communities moved toward increasingly stringent smoke-free policies over the period 2006–2009, with the greatest number of changes in policy occurring in 2008. Nearly all of the sites also reported increases in smoking cessation-related services for patients in their catchment areas in 2008–2009. However, two IS hospitals reported in 2008 that smoking cessation classes for pregnant women were not well attended. Their low attendance was attributed to patient concerns about the stigma of being labeled a "pregnant smoker." A new smoking cessation program introduced at one of the IS County Health Departments in 2009 appeared to be a promising approach to cessation. It individually tailored services for pregnant women who smoked, focused on barriers to becoming and remaining smoke free, and it followed up with clients throughout pregnancy and after birth.

Hospital Policies and Strategies Regarding Early Elective Deliveries

During the period 2007–2009, none of the IS or CS had formal, written policies that specifically prohibited elective inductions and cesarean births prior to 39 weeks, but all sites moved in the direction of policies and practices that would reduce early elective deliveries. For the IS, the biggest movement occurred in 2009, under the auspices of *HBWW* patient safety/quality improvement (QI) efforts. That year, one IS hospital sent a letter to all physicians with delivery privileges to indicate that they must adhere to ACOG guidelines. New scheduling guidelines and procedures were put into place to reduce elective inductions; and the hospital began to provide a "grade card" to physicians on their medical indication rates for elective inductions and cesareans prior to 39 weeks. The same year, the other two IS hospitals completed QI projects; one reviewed sets of medical charts for elective inductions prior to 37 weeks, while the other reviewed charts for inductions and cesarean births prior to 37 weeks.

Among the CS, some similar changes began to take place a bit earlier. In early 2006, one hospital put in place a policy prohibiting elective inductions prior to 38 weeks, and in 2007 they began to use an induction scheduling form. In 2008, a second hospital reviewed its less than 39-week elective induction and cesarean birth data. In late 2008, a third CS hospital began to track nonmedically indicated inductions and augmentations at 36–39 weeks. In the 2009 EPI, this hospital attributed the change to *HBWW*.

Across sites, nurses tended to be the first stakeholders to recognize the risk from early elective deliveries and push for policy and practice change. Physician buy-in was the biggest challenge to implementing change.

Obstetric leadership and support from other departments (Quality, Risk Management, Legal) were crucial to move change forward, as was continuous reinforcement of key messages around early elective delivery risks.

Hospital Policies Regarding Progesterone

Progesterone was not approved by the Food and Drug Administration to prevent recurrent preterm birth until after the 2007–2009 period, but it was available through compounding pharmacies. None of the IS or CS hospitals reported having a uniform policy concerning use of progesterone for this purpose.

Educational, Health, and Social Services for Pregnant Women

There was no way to validly quantify the services available for pregnant women in individual sites, or in the IS versus the CS, using the available data. However, five of six sites (i.e., all but one CS) reported increases in smoking cessation services during 2007–2009, and all three IS reported in 2008 that distribution of educational materials had increased, due (at least in part) to *HBWW*. On the other hand, the State did not carry out the planned expansion of the HANDS (Health Access Nurturing Development Services) home visitation program to multiparous women at all three IS sites. Only one IS County Health Department offered the expanded service through October 2009.

Other Environmental Factors

Several other environmental factors likely had an important impact on access to care and birth outcomes. The major economic downturn that began in late 2007 affected all of the IS and CS sites. Community members lost jobs and health insurance and in turn experienced greater stress and more health problems. Additionally, substance use among pregnant women—and a shortage of appropriate treatment services—were ongoing issues in IS and CS, particularly at one of the intervention hospitals. Finally, although many *HBWW* materials were made available in Spanish, one of the IS noted the challenge of reaching Spanish-speaking women with services, due to a lack of cross-linguistic competence among providers.

Comparison Site Exposure to *HBWW*

In the EPIs, two CS reported that staff had attended educational meetings (such as the Kentucky Perinatal Association's annual meeting) at which the latest research findings on prematurity were presented and *HBWW* messages were disseminated. One of these two CS hospitals also reported that its QI data tracking changes and its hospital leaders' increased awareness of the risks of early elective deliveries were due to *HBWW* influence. A third CS hospital reported using materials from the *HBWW* Web site in their prenatal and childbirth classes, and they made their own *HBWW* T-shirts.

PERINATAL OUTCOMES ANALYSIS

Analysis of the aggregated hospital birth certificate data revealed that from baseline to post-implementation, the largest percentage decline in the rates of singleton preterm birth and late preterm birth occurred at the IS (12.1% and 12.4%, respectively), followed by a smaller percentage decline at the CS (10.2% and 11.0%) (Table 8). Rates of singleton preterm birth and late preterm birth declined to a greater degree in both the IS and CS than they did in the rest of Kentucky (which showed declines

Table 8 Preterm and late preterm birth rates among singleton inborn live births, by hospital groups, July 2005–June 2006 and July 2009–June 2010

Hospital group	Baseline (July 2005–June 2006)		Post-implementation (July 2009–June 2010)		Change in rates	
	Count	Percent	Count	Percent	Percent change	p-value
Preterm birth (<37 completed weeks gestation)						
Intervention sites	633	15.7	512	13.8	−12.1%	0.02
Comparison sites	648	16.6	753	14.9	−10.2%	0.02
Rest of Kentucky	5255	12.2	4576	11.0	−9.8%	<0.01
Late preterm birth (34–36 completed weeks gestation)						
Intervention sites	485	12.1	395	10.6	−12.4%	0.05
Comparison sites	496	12.7	571	11.3	−11.0%	0.04
Rest of Kentucky	4061	9.4	3521	8.5	−9.6%	<0.01

of 9.8% and 9.6%, respectively). The declines in the rates of preterm birth and late preterm birth were statistically significant with the exception of late preterm births at the IS, which was slightly above the 0.05 level of significance.

Logistic regression analysis demonstrated that the odds of preterm birth declined significantly between the baseline and post-implementation time periods in the IS (OR = 0.86, 95% CI, 0.76, 0.97), the comparison sites (OR = 0.88, 95% CI, 0.78, 0.98) and the rest of KY (OR = 0.89, 95% CI 0.85, 0.93) (Table 9). Adjusting for maternal age, race, and previous preterm birth resulted in slightly lower odds ratios in the IS (OR = 0.85; 95% CI, 0.76, 0.97) and CS (OR = 0.85; 95% CI, 0.76, 0.95) for both, while the odds ratio for the rest of Kentucky was unchanged (OR = 0.89; 95% CI, 0.85, 0.93).

Given the significant findings and ecologic design of the initiative, the team also examined trends in birth data (for all births, not limited to singleton, inborn births) reported by the National Center for Health Statistics (NCHS) between 2004 and 2009 among the states in close proximity or contiguous to Kentucky. The data revealed an increase in preterm births occurring in most states and nationwide from 2004 [11] to 2006 [1] (Table 10). Kentucky (4.9%) and Ohio (6.4%) had the largest increases in preterm birth rates of the nine proximal states, contrasted with the overall national increase

Table 9 Logistic regression analysis

Unadjusted model of odds of preterm birth in the post-implementation time period compared to baseline model: Includes variable for time (post-implementation vs baseline), variable for *HBWW* participation (IS vs CS vs rest of KY), and interaction of time and *HBWW* participation to obtain time effects specific to the level of *HBWW* participation

	Odds ratio	95% Confidence interval	
Time effect in IS	0.86	0.76	0.97
Time effect in CS	0.88	0.78	0.98
Time effect in rest of KY	0.89	0.85	0.93

Adjusted model: Adds variables to adjust for differences in distributions of maternal age, maternal race, and history of previous preterm birth

	Odds ratio	95% Confidence interval	
Time effect in IS	0.85	0.75	0.96
Time effect in CS	0.85	0.76	0.95
Time effect in rest of KY	0.89	0.85	0.93

Table 10 Preterm birth rates and percent change for US, KY, and proximal states from 2004 to 2009

State	2004	2005	2006	2007	2008	2009	% Change 2004–2006	% Change 2007–2009
United States	12.5	12.7	12.8	12.7	12.3	12.2	2.4	−3.9
Kentucky	14.4	15.2	15.1	15.2	14.0	13.6	4.9	−10.5
Illinois	13.1	13.1	13.3	13.0	12.7	12.4	1.5	−4.6
Indiana	13.2	13.5	13.2	12.9	12.4	11.9	0.0	−7.8
Missouri	13.0	13.3	12.8	12.5	12.3	12.2	−1.5	−2.4
North Carolina	13.5	13.7	13.6	13.3	12.9	13.0	0.7	−2.3
Ohio	12.5	13.0	13.3	13.2	12.6	12.3	6.4	−6.8
Tennessee	14.5	14.7	14.8	14.2	13.5	13.0	2.1	−8.5
Virginia	12.1	12.3	12.0	12.1	11.3	11.4	−0.8	−5.8
West Virginia	14.0	14.4	14.0	13.9	13.7	12.9	0.0	−7.2

NCHS, 2004–2009 final birth data.

in preterm birth rates of 2.4% during this period. These increases were followed by Kentucky demonstrating the largest decrease in preterm birth rate of the nine proximal states (10.5%) in 2007 [34]–2009 [35], in contrast with the national decrease of 3.9%. In light of the fact that many of the *HBWW* educational messages were conveyed statewide, these data suggest that there may have been a positive program effect of *HBWW* across Kentucky.

CHAPTER 6

Discussion

Healthy Babies are Worth the Wait® *(HBWW)* provided an innovative model that involved clinical, public health, and community advocacy collaborations to address community-specific challenges impacting preterm birth. The initiative sought to determine if the rising rate of preterm birth could be reversed by a program of bundling evidence-based interventions and developing partnerships among key stakeholders. The result of this pilot, evaluated employing both qualitative and quantitative methods, demonstrated the success of the *HBWW* initiative to reduce preterm births in Kentucky and provides evidence for an effective intervention that can be applied in other communities.

During the 3-year period of the initiative, a statistically significant decrease in preterm births occurred not only in the intervention sites (IS) but also in the comparison sites (CS)—OR 0.85 in both sets of sites—and the rest of Kentucky (OR 0.89). The hospitals in the rest of Kentucky were not included in the formal study design. The decrease in preterm birth rates in the nonintervention hospitals, while not anticipated, is not surprising considering the widespread dissemination of prematurity prevention-related information that occurred statewide. The intervention's lead implementation agencies, viz. Kentucky Department for Public Health and March of Dimes, had substantial reach not only to the IS, but also to the CS and the rest of Kentucky; and a substantial program component involved nonexclusive access to mass media messages and educational conferences.

The 2007–2009 interval of the intervention also coincided with statewide efforts by the Kentucky Perinatal Association to educate perinatal providers about late preterm birth, thus publicizing *HBWW* and further "contaminating" nonintervention providers. In addition, a nationwide prematurity awareness campaign by March of Dimes mirrored many aspects of the *HBWW* intervention. Thus, there was a high probability that "unintended" exposure occurred in the CS and the rest of Kentucky to the interventions intended by study design solely for the IS groups.

Consumer and provider surveys conducted in 2009 provide evidence that the CS had been exposed to *HBWW* messages and materials. Among consumers, 8% reported having heard of *HBWW* and 13% reported seeing a signature *HBWW* educational card on fetal brain development. Forty-four percent of providers reported having heard of *HBWW*, 11% reported having used at least one of eight identified *HBWW* materials, 10% reported having participated in *HBWW* continuing education activities, and 7% reported having used the *HBWW* Web site.

Information from the Environmental and Policy Interviews (EPI) provided further evidence for exposure of the CS hospitals to *HBWW* interventions. Two CS hospitals stated that their staff had attended educational meetings on prematurity where *HBWW* messages were disseminated. One hospital reported that its quality improvement (QI) data tracking changes and its hospital leaders' increased awareness of the risks of early elective deliveries were due to *HBWW* influence. A third CS hospital reported using materials from the *HBWW* Web site in their prenatal and childbirth classes, and creating their own *HBWW* T-shirts.

The unintended exposure of *HBWW* messaging and educational content throughout the state of Kentucky is evident from the activities described above and can potentially account for the widespread success of the initiative to reduce preterm birth, not only in the IS, but in the CS and the rest of Kentucky as well. To support the premise of a statewide impact of *HBWW* in Kentucky, we also examined the trends in preterm birth for all births in the US and among the states in close proximity or contiguous to Kentucky for the calendar years 2004–2009. Analysis of the percent change in preterm birth occurring between 2007 and 2009, during the initiative's implementation, confirmed that Kentucky had the largest percent decrease among the proximal states of 10.5% compared to a 3.9% decrease in preterm birth that occurred nationally. The second largest percent decrease of 8.5% occurred in Tennessee and is notably reduced from the 10.5% decline that occurred in Kentucky. These findings provide further evidence for the successful statewide impact of *HBWW* as a result of "contamination" throughout Kentucky in contrast to the lesser declines in preterm births that occurred nationally and in neighboring states.

An examination of the literature provided scant evidence for similar initiatives that fostered collaborative relationships between medical, public health, and community partners; implemented QI processes; applied bundled interventions; and effected service integration to achieve efficiencies in patient care and improve preterm birth outcomes. One study by Leveno and colleagues, published in 2009, describes their public health approach to the provision of clinical care, emphasizing access and the removal of barriers to service [36]. They reported a dramatic reduction in preterm birth among minority women from 10.4% to 4.9% between 1995 and 2002 at Parkland Memorial Hospital in Dallas, Texas. The researchers described their prenatal care services as "one component of a comprehensive and orchestrated public health-care system that is community-based" and that specifically targets minority populations of pregnant women. Through their service approach, Parkland increased the percent of patients receiving prenatal care to 98% in 2006 (from 88% in 1988) and achieved rates of preterm birth below the national average in an inner city, high-risk prenatal population consisting of 70% Hispanic and 20% African-American women. The success achieved at Parkland Memorial Hospital in reducing preterm birth through the application of a service model with program elements similar to *HBWW* provides further evidence in support of *HBWW*'s potential to effect a positive change in perinatal birth outcomes and reduce preterm birth when rigorously applied.

Significant strengths of the initiative were the power of the collaborative relationships established between the IS hospitals and their local health departments, and the IS hospitals' activities to leverage cross-site learning opportunities. Each IS enhanced or developed new procedures for communication between the hospitals and local health departments that enabled improvements to patient counseling and the coordination of referrals for services. A collaboration at one IS resulted in the sharing of staff by stationing a health department nurse at the hospital clinic to provide education to women during their prenatal visit and arrange access to programs such as WIC (Special Supplemental Nutrition Program for Women, Infants and Children) and home visitation case management services. The cooperation between the IS hospitals was most notable in their QI activities, where tools and trainings were shared across hospital institutions.

A key lesson in the implementation of *HBWW* was the importance of strong leadership from within the hospital-based and public health nursing staff to serve as champions and the driving impetus behind the initiative. Physician champions were equally important; however, nurses provided the largest contingent of unified manpower and influence to generate change within their institutions. Several recommendations were made by the participating sites as guidance for future *HBWW* initiatives. They identified the importance of building strong partnerships and establishing trust, and engaging diverse partners in team development, including hospitals, clinics, and health departments as well as schools, civic organizations, consumer representatives, and programs such as Early Head Start and Healthy Start.

A formal orientation of the staff at new sites was also emphasized in order to leverage experiences and best practices from experienced sites through ongoing meetings, conference calls, and mentorship. Providing outcome data in real time to demonstrate the impact of the institutional and team efforts was also stressed.

The 12% decrease in preterm birth from baseline occurred despite external challenges encountered during the implementation of the *HBWW* initiative. In particular, the initiative took place during a national recession that had severe economic impact across the state. This was reflected in provider-reported food shortages, financial barriers to health-care access based on loss of insurance coverage, and a rise in substance abuse with few treatment programs available for pregnant women. That a substantial decrease in preterm births occurred despite these challenges demonstrates the robustness of the intervention strategies.

Challenges to the implementation of the initiative were encountered that required innovative approaches to resolve in many instances. Both providers and pregnant women found ways to circumvent the established hospital smoke-free policy. Community complaints of hospital personnel and patients smoking on adjacent properties resulted in one of the IS hospitals creating a "Smoke Shack" which was not a desirable or positive solution. Smoking cessation classes for pregnant women were initially not well attended, due to the stigma of being identified as a "pregnant smoker." An alternative program, Giving Infants and Families a Tobacco-Free Start, was established at one of the IS through a collaboration with the University of Kentucky Department of Nursing and the Kentucky Department for Public Health. It provided individualized tailoring of services to pregnant smokers through a bundling of research-based interventions that included individu-alized motivational interviewing, 5A's counseling, Tobacco Quit Line referrals, and social support screening. The program successfully generated increased patient participation and achieved higher quit rates among pregnant smokers.

An important *HBWW* service activity was to have providers make appropriate patient referrals to a social worker who would link pregnant women to prenatal classes, home visitation services, drug-use related and smoking cessation services, and domestic violence services. An incentive for obstetrical providers to refer patients to the social worker was created at one IS hospital that experienced a low referral rate. The hospital began giving "report cards" to the providers on their referrals and this feedback successfully catalyzed a friendly competition among providers to make needed referrals.

In the area of resource demand, an unanticipated investment of additional time was needed by dedicated *HBWW* program personnel to support internal communications and provide continuous technical assistance. These activities significantly enhanced the overall effectiveness of the initiative by providing individualized guidance to the IS on strategies to engage their hospital administration and improve risk management and QI activities.

Ecologic study designs contain inherent limitations, particularly the risk of contamination; yet to evaluate programmatic impact they often offer the most appropriate and cost-effective approach, particularly when there is a dynamic environment and time issues are paramount. Extensive contamination was anticipated between the IS and CS, despite their separate geographic locations, due to the intensive statewide continuing education activities, inclusion of many key perinatal leaders in the state in program activities, wide coverage of the media messages, and public Web site access to initiative materials. Broad exposure to the initiative also occurred through presentations at multiple national forums.

Given the evaluation design, the effects of the specific program components were not able to be considered individually. Moreover, the relative homogeneity of the population precludes generalization of the findings to more diverse communities, particularly in high-risk inner-city neighborhoods.

HBWW achieved a statistically significant reduction in preterm birth rates over 60 months, not only in the IS but also in the CS and the remainder of Kentucky. This initiative provides evidence that an effective collaborative model can be devised involving clinical and public health professionals, hospitals, and community organizations. Initiation of similar, tailored initiatives responsive to local risk factors and resources in other states and communities could decrease preterm birth in these venues. Interventions should be identified based on the needs of the community taking into account the population demographics, the prevalence of relevant medical conditions, behavioral and social factors, access to care, provider practice patterns, and available resources in the community and for the intervention. Of potential economic and pragmatic benefit, implementation at a single site can be expected to have salutary effects beyond the immediate venue.

Lessons learned from the *HBWW* pilot in Kentucky has guided dissemination of the initiative to the three CS and two additional communities in Kentucky (2010–2011). Three communities in Houston, Texas; and two communities in Newark, New Jersey were added in 2012. During 2014, Kentucky expanded *HBWW* to a ninth community, Kansas adopted the initiative in seven communities, and the program was launched in four sites in Western New York for a total of 26 operating communities in five states at the end of 2014. Because of the success of this model in bringing together clinical, public health, and community advocates as partners, the March of Dimes has developed a formal dissemination plan and will continue its commitment to replicate *HBWW* in additional states and communities across the US.

APPENDIX A
HBWW LOGIC MODEL

Principal Targets: Pregnant women and the hospitals and hospital-based health-care providers who serve them in three communities in Kentucky (KY)

Inputs	Strategies/Activities	Outputs	Short-Term Outcomes	Mid-Term Outcomes	Long-Term Outcomes
JJPI	*For pregnant women:* Home visitation, KY's Tobacco Quit Line, folic acid, prenatal care education, dental services, substance use treatment, group practice, and resources	*For pregnant women:* Home visits made; other services provided; materials disseminated	*For pregnant women:* Changes in knowledge and attitudes re: preterm birth circumstances and modifiable risk factors, with a focus on late preterm (e.g., why important to go to term, birth options, smoking, folic acid, etc.)	*For pregnant women, providers, communities:* Increase in preterm birth prevention-related behaviors (e.g., including but not limited to reduction of elective inductions and cesarean sections prior to 39 weeks' gestation)	Reduction in singleton preterm birth rate (15%)
MOD					
KY Dept. for Public Health					Reduction in preterm birth-related hospital charges
Hospital sites	*For perinatal providers:* Continuing medical education; provision of current research, tools, and patient education materials	*For perinatal providers:* Grand rounds and related trainings delivered and made available via the Internet; tools, articles, and other materials provided; referral systems enhanced	*For perinatal providers:* Changes in knowledge and attitudes re: ACOG guidelines on elective inductions and cesareans and preterm birth prevention (especially late preterm birth); and related AAP and AWHONN guidelines		
Pregnant women					
Local KY communities			Increased knowledge and use of referral systems for public health services for high-risk women		Reduction in neonatal morbidity associated with preterm birth
Communications experts					
Health-care leadership	*For communities:* Media campaign, educational presentations, provision of educational materials	*For communities:* Message points developed; radio/TV spots broadcast; community tool kits developed/used	*For communities:* Changes in knowledge and attitudes re: family planning, preconception health, and importance of preterm birth prevention		
Existing ACOG, AWHONN, and AAP evidence-based practice guidelines	*For hospitals:* Provision of materials on good institutional practices, including data collection and reporting	*For hospitals:* Materials provided	*For hospitals:* Changes in policies and practices concerning quality assurance, patient safety, smoke-free campuses, data collection and reporting, and provision of relevant educational information to women of child-bearing age		Replication of the intervention model in other vulnerable states

Notes: AAP=American Academy of Pediatrics; ACOG=American College of Obstetricians and Gynecologists; AWHONN=Association of Women's Health, Obstetric and Neonatal Nurses; JJPI=Johnson & Johnson Pediatric Institute, L.L.C.; MOD=March of Dimes.

APPENDIX B
HBWW ORGANIZATIONAL STRUCTURE

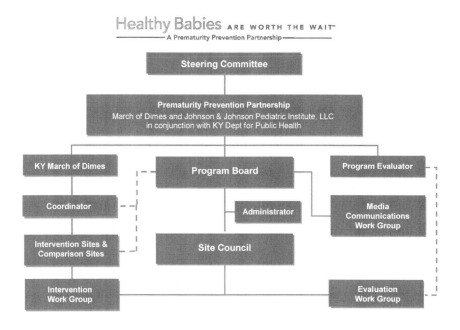

APPENDIX C
COMMUNITY EDUCATION AND MARKETING MATERIALS

HBWW EDUCATIONAL MATERIALS

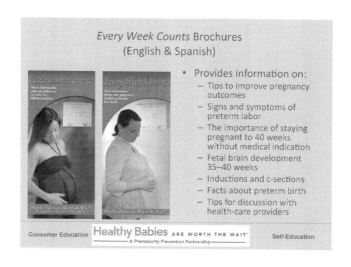

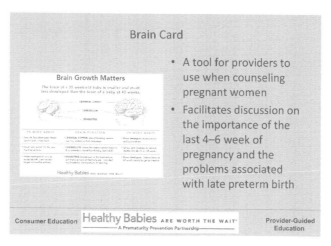

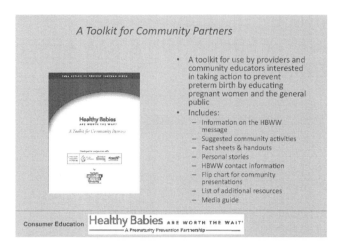

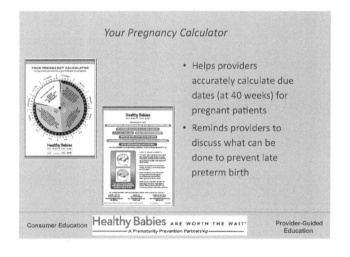

HBWW MARKETING MATERIALS

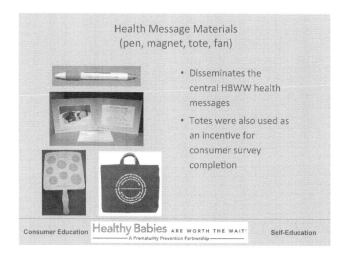

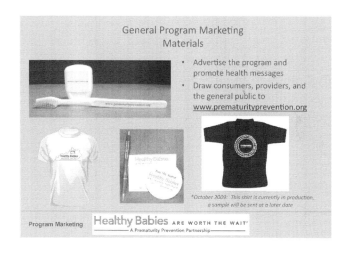

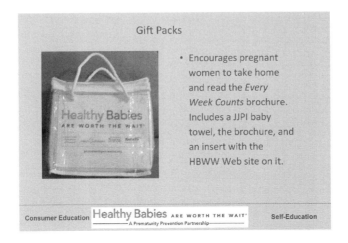

LIST OF PRODUCTS CREATED FOR *HBWW*: INCLUDES MESSAGING

Appointment reminder cards
Brain card
Every Week Counts brochure (English & Spanish)
Johnson & Johnson Pediatric Institute, LLC Pregnancy Diary Computer Program Reminder Card
Johnson & Johnson Pediatric Institute Pregnancy, LLC Diary Computer Program Reminder Insert/Flyer
Magnets
Map T-shirt
Message fan
Message pens
Message totes
Message T-shirt
Pregnancy wheel
A toolkit for community partners (with tote bag and flipchart)

LIST OF PRODUCTS CREATED FOR *HBWW*: LOGO AND/OR WEB SITE ONLY

"Ask me about HBWW" buttons
Dental floss
Purple pens
Sticky notes
Toothbrushes
Logo fan
Stationary: letterhead (2 versions), envelopes, notecards

APPENDIX D
HBWW PROCESS DATA REPORT

DATA FROM APRIL 2007 THROUGH NOVEMBER 2009 (UNLESS OTHERWISE NOTED)

Activities	Numbers	
Conference presentations	11 national; 26 state; 29 local 1 KY Prematurity Summit (featured a *HBWW* panel)	
Provider education	18 Grand Rounds presentations (~480 attendees) 27 *HBWW* training sessions (~1200 attendees) 131 peer-reviewed articles and news articles (*from October 2008*)	

HBWW materials distributed to intervention sites (*from August 2007*)	7275 English and 1590 Spanish brochures
	216 brain cards
	984 toothbrushes
	1122 dental floss
	933 message pens
	3120 magnets
	1320 tote bags
	2200 Program Newsletter
		525 Patient Newsletter
		194 map T-shirts:
		715 Ask Me About *HBWW* pins
		Online pregnancy diary displays:
		    90 counter cards
		    3000 take-away cards
		897 message fans
		115 message T-shirts

Program meetings	67 conference calls (combinations of the executive program board, intervention sites, and local health departments) 5 site council meetings	
Web site usage	65,951 total visitors (*from January 2008*) First quarter 2008: 5851 Second quarter 2008: 9300 Third quarter 2008: 6711 Fourth quarter 2008: 7126 First quarter 2009: 7346 Second quarter 2009: 8006 Third quarter 2009: 11,061	122 downloads of pregnancy diary computer program (*from December 2008*) 268 Facebook group members (*from March 2009, count as of 12/9/09*) 65 Twitter followers 113 Twitter messages (*from June 2009, counts as of 12/9/09*)
Press coverage (*from August 2007*)	65 instances of original press coverage on *HBWW*	

APPENDIX E
CONSUMER AND PROVIDER SURVEYS
(FOLLOW-UP VERSIONS)

Healthy Babies are Worth the Wait® **- Survey for** <u>**Pregnant Women**</u>

*This survey is voluntary and anonymous--please **do not** put your name on it. Your answers will help us create better materials and programs for pregnant women. Skip any questions that you do not want to answer.*

1. How many weeks pregnant are you? _____ Weeks
   If you do not know the number of weeks, please tell us how many
   months. _____ Months ☐ Not sure

2. How many prenatal visits have you had in <u>this</u> pregnancy?
   ☐ This is my first visit or _____ prenatal visits so far

3. How many times have you given birth before? _____ Times
   ☐ Never given birth before *(skip to question 4)*
   a. How many of these babies were born preterm
      ("premature" or "early")?
      _____ Babies ☐ None
   b. How many were born weighing 5 ½ pounds or less?
      _____ Babies ☐ None
   c. How many were delivered by cesarean delivery (or "c-section")?
      _____ Babies ☐ None
   d. How many times did your health-care provider (doctor, nurse, etc.)
      have to start (induce) labor?
      _____ Times ☐ None
   e. Have you ever had twins or triplets? ☐ Yes ☐ No
   f. What was the date of your last delivery? _____ At what
      hospital? _____

4. Why do you think babies are sometimes born too early (prematurely)?
   _____
   _____

5. How long is a term pregnancy?
   ☐ 28–31 weeks ☐ 35–36 weeks ☐ Not sure
   ☐ 32–34 weeks ☐ 37–41 weeks ☐ Other: _____

6. Has a health-care provider (doctor, nurse, etc.) ever talked with you about the signs and symptoms of having labor too early (preterm labor)? ☐Yes ☐No ☐Not sure

7. If your baby was born three weeks before the due date, would this be a problem? *(check only one)*
   ☐Not really a problem ☐Somewhat serious ☐Serious
   ☐Very serious

8. How serious of a problem is preterm or premature birth in your community?
   ☐Very serious ☐Somewhat serious ☐Not at all serious
   ☐Not sure

9. Has a doctor recommended a cesarean delivery for your current pregnancy? ☐Yes ☐No ☐Not sure
   a. If yes, what was the reason? _____
   _____ ☐Not sure

10. In a pregnancy without problems for the mother or baby, please tell us if you think the following statements are true or false:
    a. Cesarean delivery is usually *safer* than vaginal delivery.
       ☐True ☐False ☐Not Sure
    b. It is a good idea to schedule the date of delivery for convenience at 35 or 36 weeks.
       ☐True ☐False ☐Not Sure
    c. Women have more problems later in life if they have had vaginal (instead of cesarean) births.
       ☐True ☐False ☐Not Sure

11. Before you got pregnant, how often did you take a multivitamin or folic acid supplement each day?
    ☐Daily ☐Weekly ☐A few times each month ☐Never

12. During this pregnancy, are you taking a prenatal vitamin or multivitamin each day?
    ☐Daily ☐Weekly ☐A few times each month ☐Never

13. If your health-care provider (doctor, nurse, etc.) recommended that you take a multivitamin every day, how likely would you be to follow their advice?
    ☐Very likely ☐Somewhat likely ☐Not too likely
    ☐Not at all likely

14. How would you describe your health?
    ☐Excellent ☐Good ☐Fair ☐Poor

15. Do you need any of the following services? *(Check as many as you need.)*
    - ☐ Money to buy food, food stamps, or WIC vouchers
    - ☐ Help to quit smoking
    - ☐ Help with an alcohol or drug problem
    - ☐ Help reducing violence in your home
    - ☐ More information on how to have a safe, healthy pregnancy
    - ☐ Childcare for current children
    - ☐ Other:_____

16. During this pregnancy, how many days a week do you usually feel blue, stressed, or depressed? _____ days

    a. Have you received any counseling or treatment for these feelings during this pregnancy?   ☐ Yes ☐ No

17. Have you visited a dentist in the past year? ☐ Yes   ☐ No

18. How would you describe your teeth?
    - ☐ Good condition ☐ Need some work ☐ Poor condition
    - ☐ Other _____

19. During this pregnancy, has a health-care provider (doctor, nurse, etc.) spoken to you about progesterone (also known as "Gestiva")?
    ☐ Yes ☐ No ☐ Not sure

The next few questions are about your tobacco and alcohol use.

20. In the month before you became pregnant, how many cigarettes did you usually smoke? _____ per day OR _____ per week   ☐ Did not smoke

21. During this pregnancy, how many cigarettes do you usually smoke? _____ per day OR _____ per week   ☐ Do not smoke

22. During this pregnancy, has a health-care provider (doctor, nurse, etc.) told you not to smoke or to stop smoking? ☐ Yes   ☐ No

23. In the month before you became pregnant, how many beers, glasses of wine, or other alcoholic drinks did you usually drink? _____ drinks per day OR _____ drinks per week   ☐ Did not drink

24. During this pregnancy, how many beers, glasses of wine, or other alcoholic drinks do you usually drink? _____ drinks per day OR _____ drinks per week   ☐ Do not drink

25. During this pregnancy, has a health-care provider (doctor, nurse, etc.) told you not to drink alcohol?   ☐Yes ☐No

26. Did you change any of your behaviors (such as smoking, drug use, drinking, diet, or exercise) since you found out that you were pregnant?   ☐Yes ☐No
   a. If yes, describe the change(s) you have made:
   _____

27. Was this pregnancy planned?   ☐Yes ☐No

28. During the year before this pregnancy, did you see a health-care provider (doctor, nurse, etc.)?   ☐Yes ☐No
   a. If yes, what was the reason?
   _____

29. During this pregnancy, how many ultrasounds (sound pictures) have you had?   _____ultrasounds   ☐Not sure

30. How much do you know about the following pregnancy topics?
   a. Good eating habits
      ☐A lot   ☐A little   ☐Nothing at all
   b. Exercising during pregnancy
      ☐A lot   ☐A little   ☐Nothing at all
   c. Smoking during pregnancy
      ☐A lot   ☐A little   ☐Nothing at all
   d. Dealing with stress and depression
      ☐A lot   ☐A little   ☐Nothing at all
   e. Preventing infections
      ☐A lot   ☐A little   ☐Nothing at all
   f. Delivery method options (such as c-section vs vaginal birth and pain management options)
      ☐A lot   ☐A little   ☐Nothing at all

31. During this pregnancy, has your health-care provider (doctor, nurse, etc.) talked to you about any of the following services, and have you used them? (Check all that apply for each lettered item.)
   a. Kentucky QuitLine (to stop smoking)
      ☐Yes talked about   ☐Yes used the service   ☐No
   b. HANDS (home visitation program)
      ☐Yes talked about   ☐Yes used the service   ☐No
   c. Domestic violence services
      ☐Yes talked about   ☐Yes used the service   ☐No
   d. Dental services
      ☐Yes talked about   ☐Yes used the service   ☐No

32. Have you heard of *Healthy Babies are Worth the Wait*® (HBWW)?

☐Yes   ☐No   ☐Not sure

If yes: a. What is *HBWW* about? _____

_____

b. Where did you hear about *Healthy Babies are Worth the Wait*®? *(Check all that apply.)*

☐From my doctor, midwife, or nurse
☐Brochure, poster
☐Community presentation
☐Health fair
☐Community information booth
☐Preterm birth prevention handouts/fact sheets
☐Media coverage (TV, newspaper, or radio)
☐Advertisement (magazine, billboard)
☐T-shirt or button with *HBWW*
☐Pre-admission testing
☐Childbirth classes
☐Other: _____

33. During your current pregnancy, has your health-care provider (doctor, nurse, etc.) showed you a card with pictures of how a baby's brain develops during pregnancy?

☐Yes   ☐No   ☐Not sure

34. Which of the following items have you received that have *Healthy Babies are Worth the Wait*® or *www.prematurityprevention.org* on the item? *(Check all that apply.)*

☐ "Every Week Counts" brochure
☐ Pen with changing health messages
☐ Tote bag
☐ Pregnancy diary
☐ Dental floss or toothbrush
☐ Magnetic picture frame
☐ Other items, please list: _____
☐ I did not receive any items with *Healthy Babies are Worth the Wait*® or *www.prematurityprevention.org*

35. How many times have you used the Web site *www.prematurityprevention.org*? ____ times or ☐ Never used it

a. If you have ever used it, how would you describe it?

☐Excellent   ☐Very good   ☐Good   ☐Fair   ☐Poor
☐Not sure

These last few questions tell us a little about you.

36. What is your marital status?

☐Married   ☐Single
☐Single and living with partner ☐Other _____

37. Please describe the people living in your household.
   a. Number of adults living with you:
      ☐None   ☐1   ☐2   ☐3 or more
   b. Number of cigarette smokers:
      ☐None   ☐1   ☐2   ☐3 or more
   c. Number of children (18 years or younger):
      ☐None   ☐1   ☐2   ☐3 or more

38. Does the father of your baby live in your home?
      ☐Yes   ☐No

39. How old are you? _____ years old

40. How much schooling have you had?
     ☐Up to 8th grade
     ☐9th–12th grade, no diploma
     ☐High school graduate or GED
     ☐Some college credit but no degree
     ☐Associate's degree
     ☐Bachelor's degree or higher

41. Are you of Hispanic (Spanish/Latina) origin? ☐Yes   ☐No

42. Please check off the race that you most consider yourself to be.
   (Check only one.)
     ☐White   ☐Asian or Pacific Islander ☐Other _____
     ☐Black or African American   ☐Native American

43. How are you paying for this visit today?
     ☐Private insurance   ☐Self-pay (cash, check, or credit card)
     ☐Medicaid   ☐Other _____

Thank you for completing this survey. Please return it to the front desk.

If you have questions, contact (name) at (email) or (phone).

2009 Provider Survey
Healthy Babies are Worth the Wait®

This voluntary, anonymous survey should be completed by providers (physicians, nurse midwives, nurses) who currently care for <u>pregnant women</u>. ***Please answer <u>both pages</u>***.

1. What do you think are two key factors that led to the increase in preterm birth rates in the past decade?

    (1)_____

    (2)_____

    ☐Not sure

2. Do you/does your facility currently offer, recommend or refer eligible pregnant patients for progesterone therapy for the prevention of recurrent preterm birth?
    ☐Yes: (☐Offer ☐Recommend ☐Refer) ☐No ☐Not sure

3. Please check the box that best describes how concerned you are about each of the following issues related to progesterone therapy during pregnancy to prevent recurrent preterm birth:

	Very Concerned	Somewhat Concerned	Not Concerned		Very Concerned	Somewhat Concerned	Not Concerned
a. Safety				f. Not FDA approved			
b. Efficacy				g. Liability			
c. Not reimbursed by Medicaid				h. Patients don't want injections			
d. Not covered by insurance				i. Need for more data			
e. Not easily available				j. Long-term fetal/ neonatal effects			

4. How often do your <u>patients request</u> progesterone to prevent recurrent preterm birth?
    ☐Frequently   ☐Infrequently   ☐Never   ☐Not sure

5. Please check the box that best describes what you think of the following factors as contributors to preterm labor/birth:

	Very Important	Somewhat Important	Not Important	Not Sure		Very Important	Somewhat Important	Not Important	Not Sure
a. Smoking while pregnant					g. Periodontal disease				
b. Second-hand smoke					h. Maternal stress				
c. Any/occasional alcohol intake					i. STDs				
d. Heavy alcohol intake					j. BV				
e. Illicit drugs (meth, cocaine)					k. UTI				
f. Domestic violence					l. Maternal obesity				

6. Which statements reflect your beliefs about smoking cessation during pregnancy? *(Check all that apply.)*
   ☐ Very important ☐ Not important ☐ Costs are a barrier
   ☐ No services available for pregnant women
   ☐ Women won't participate
   ☐ Other:_____

   a. How often do you/does your facility provide or refer pregnant women for smoking cessation?
   ☐ Frequently ☐ Infrequently ☐ Never
   ☐ Not sure

   b. Which smoking cessation resources do you use?
   (Check all that apply.)
   ☐ Programs at our facility ☐ 5As
   ☐ KY QUIT LINE ☐ National Quitline ☐ Local Health Dept
   ☐ Other:_____ ☐ Not sure

7. Based on ACOG practice guidelines, elective inductions (no medical or obstetrical indication) should not be done until at least _____ weeks of gestation ☐ Not sure

8. How concerned are you about the increasing rates of late preterm birth (34–36 weeks)?
   ☐ Very concerned ☐ Somewhat concerned
   ☐ Not concerned ☐ Not sure

9. What types of QI/patient safety strategies or protocols do you think would be appropriate to consider for:
   a. primary elective cesarean delivery: _____
   _____
   b. elective inductions:_____
   _____
   c. preterm birth:_____
   _____

10. Please check the box that best describes your reaction to the following statements about changes in ob practice.

	Strongly Agree	Agree	Disagree	Strongly Disagree	Not Sure
a. Increases in ART and infertility Rx are causing most of the increase in preterm births					
b. Scheduling delivery at 34–36 wks is usually not a problem for low risk singleton pregnancies					
c. Cesarean delivery on maternal request is usually not a problem for low risk pregnancies					
d. More patients want to schedule the date of their delivery					
e. Increasing maternal age is causing most of the increase in preterm births					
f. Malpractice issues have a major impact on obstetrical clinical practice					
g. VBAC should no longer be offered as an option for delivery					
h. Most women would rather have a cesarean than a vaginal delivery					

11. Assuming that you or your spouse has a nulliparous, uncomplicated, low risk, singleton, cephalic pregnancy, and you can determine how the delivery will occur:

   a. What mode of delivery would you want/recommend?
      ☐ Vaginal delivery   ☐ Elective cesarean delivery
   b. At what week of gestation would you want the delivery to occur? _____ weeks
   c. Would you consider an elective induction?
      ☐ Yes   ☐ No   ☐ Not sure

12. Have you heard of the initiative *Healthy Babies are Worth the Wait*®
    (HBWW)? ☐Yes ☐No ☐Not sure
    If yes: a. What is *HBWW* about?_____
             b. How would you describe its impact?
                 ☐Effective   ☐Not effective   ☐Not sure
             c. What did you like most about *HBWW*?_____
                 _____
             d. What did you not like about *HBWW*?_____
                 _____

13. Which of the following *Healthy Babies are Worth the Wait*® materials
    have you used? (*Check all that apply.*)
    ☐*Every Week Counts* brochure
    ☐Fetal Brain Development Card
    ☐Fax Referral Form
    ☐Prematurity Prevention ToolKit
    ☐*HBWW* pregnancy diary
    ☐*HBWW* pens
    ☐*HBWW* dental floss/toothbrush
    ☐*www.prematurityprevention.org*
    ☐None of the above

14. Did you participate in any *Healthy Babies are Worth the Wait*®
    continuing education? ☐Yes ☐No ☐Not sure

    If yes: a. Check the types of continuing ed you used:
                 ☐Grand rounds   ☐Conference
                 ☐Webinars   ☐Articles
             b. How would you describe its impact?
                 ☐Effective   ☐Not effective   ☐Not sure

15. About how many times have you used the Web site
    www.prematurityprevention.org? ___ times or ☐Never used it

    If ≥1 time: a. How would you describe it?
    ☐Excellent ☐Very good ☐Good ☐Fair
    ☐Poor ☐Not sure

16. What do you think is the most important intervention to prevent
    preterm birth?_____

The next questions are about you and your work.

17. Are you a: ☐MD   ☐DO   ☐Midwife ☐PA   ☐NP   ☐Nurse
    ☐Other:_____

    a. What is your specialty?   ☐Ob/Gyn ☐FM   ☐MFM ☐REI
    ☐Other:_____

18. Your clinical practice/practice setting is best described as:
    ☐Solo
    ☐Academic med center/faculty practice
    ☐Hospital based (non-academic)
    ☐Single/multi-specialty group
    ☐HMO (staff model)
    ☐Other:_____

19. In what area(s) do you primarily practice?
    ☐Urban   ☐Suburban
    ☐Rural   ☐Other:_____
    a. In which counties do you practice in Kentucky?_____
    _____

20. How long have you been caring for pregnant women? _____
    years

21. Annually about how many pregnant patients do you provide care
    for? _____pregnant patients

22. What is:
    a. Your age category: ☐<40 ☐40–49 ☐50–59 ☐60–69 ☐70+
    b. Your gender: ☐Male   ☐Female

Comments/suggestions:_____
_____

APPENDIX F
ANNUAL PERINATAL SYSTEMS SURVEY

Annual Perinatal Systems Survey

The purpose of this survey is to collect basic information about your hospital's perinatal systems and about some structural and environmental factors that may influence preterm birth rates in your hospital catchment area.

1. How many OB providers have admitting privileges at your hospital?
   Total #:_____

   a. Specify # by specialty:

   OB/GYN: _____   MFM: _____   Midwives: _____
   Family Med: _____   Other: _____

2. Annually your facility has about how many:

   a. Maternal transports in? _____

   b. Maternal transports out? _____

3. How many prenatal clinics/physician offices provide prenatal care for women who deliver at your hospital? _____

   a. Please provide the requested information for each prenatal care facility associated with your facility on the attached sheet: *name of facility, number of patients seen annually, number of visits annually, proportion on Medicaid.*

4. Which of the following services (e.g., classes, medical care, social services, etc.) are available <u>for pregnant women</u> in your hospital catchment area?

	<u>Yes</u>	<u>No</u>	<u>Comments/describe</u>
a. Prenatal classes	___	___	_____
b. Smoking cessation	___	___	_____
c. Dental care/oral health	___	___	_____
d. Domestic violence	___	___	_____
e. Substance abuse	___	___	_____
f. Weight management	___	___	_____
g. Stress management	___	___	_____
h. Other: _____			

5. Does your facility offer regular continuing education (CE) venues for your obstetrical providers?

__ Yes __ No __ Not sure

a. IF YES: How often, what type of CE (general medical, nursing, or OB/GYN-specific), and in what format (e.g., grand rounds, web-based, etc.)?

How often: _____ Type of CE: _____

Format: _____

6. Is your hospital equipped for telemedicine? __ Yes __ No __ Not sure

7. Please describe your hospital's policy or policies regarding the following (and provide copies of these policies, if available):

a. Elective induction _____

b. Elective primary cesarean delivery _____

c. Newborn admission to the NICU _____

d. Smoking on the hospital campus _____

e. Provision of educational materials to women of child-bearing age

_____

f. Use of progesterone to prevent recurrent preterm birth

_____

8. Does your OB service at your hospital have a patient safety/quality improvement program?

__ Yes __ No __ Not sure

a. IF YES: What strategies or protocols are in place concerning the following:

Elective induction _____

Elective primary cesarean delivery _____

9. Please fill in the following information about your hospital, for 2007 and for 2008:

	2007	2008
a. Number of live births		
b. Overall preterm birth rate		
c. Cesarean rate		
d. Induction rate		
e. % of deliveries that were scheduled		

Information about Prenatal Care Sites with Patients that Deliver at Your Institution

Prenatal Care Site	# Pts Seen Annually	# Visits Annually	% on Medicaid

APPENDIX G
ENVIRONMENT AND POLICY INTERVIEW

ANNUAL ENVIRONMENT AND POLICY INTERVIEW WITH INTERVENTION SITES

This interview accompanies the Annual Perinatal Systems Survey that your hospital is completing. The purpose of this interview is to collect more in-depth information on structural and environmental factors that may influence preterm birth rates in your hospital catchment area.

1. What changes in hospital policy, if any, have been implemented during the past 12 months concerning
    (a) Elective induction
    (b) Elective primary cesarean delivery or cesarean delivery on maternal request
    (c) Newborn admission to the NICU
    (d) Smoking on the hospital campus
    (e) Provision of educational materials to women of child-bearing age
    (f) Provision of progesterone to prevent recurrent preterm birth
    (g) Other policies that may affect preterm birth rates

2. If any changes have taken place, per Question #1:
    (a) When did the changes occur?
    (b) What were the reasons for these changes?
    (c) What challenges have been faced in implementing the policy changes, and how have they been addressed?
    (d) What successes related to the policy changes should be noted?

3. What changes in patient safety/quality improvement program strategies or protocols, if any, have been implemented in the past 12 months concerning
    (a) Elective induction
    (b) Elective primary cesarean delivery

4. If any changes have taken place, per Question #3:
    (a) What were the reasons for these changes?

(b) What challenges have been faced in implementing the policy changes, and how have they been addressed?

(c) What successes related to the policy changes should be noted?

5. What changes, if any, have taken place in the past 12 months with respect to the availability of the following services (e.g., classes, medical care, social services, etc.) for pregnant women in your hospital's catchment area? (Please describe the nature, causes, and effects of these changes.)

(a) Prenatal classes

(b) Smoking cessation

(c) Dental care/oral health

(d) Domestic violence

(e) Substance abuse

(f) Weight management

(g) Stress management

(h) Other

6. What other factors, if any, have changed in the past 12 months that are affecting or will affect perinatal outcomes? Please describe. For example:

(a) Changes in national, state, or local legislation/policies (e.g., concerning insurance coverage, reimbursable medical services, prosecution of substance-using women, smoking in public places, etc.).

(b) Changes in funding for implementation or enforcement of relevant legislation/policies.

(c) Changes in the availability of health-care services outside your hospital (e.g., closing or opening of other clinics or hospitals, expansion or reduction of their hours or services, etc.).

(d) Initiation or conclusion of community-wide or targeted health prevention or promotion campaigns (e.g., media campaigns)

(e) Other factors not discussed already in the interview.

7. Please describe any media coverage over the past 12 months on the Healthy Babies are Worth the Wait® (*HBWW*) initiative (e.g., timing, medium, message, target audience) that you are aware of in your hospital's catchment area.

8. What aspects or characteristics of the prenatal care sites and/or population in your hospital's catchment area have presented the greatest challenges to the *HBWW* initiative? What strategies have been used to address them, and how successful have they been?

9. What do you perceive to have been the *most valuable or impactful aspect* of your hospital's participation in the *HBWW* initiative?

10. What do you perceive to have been the *most challenging aspect* of your hospital's participation in the *HBWW* initiative?

11. Over the course of *HBWW*, what factors external to the *HBWW* project do you feel have had the greatest impact on project implementation and/or outcomes?

ANNUAL ENVIRONMENT AND POLICY INTERVIEW WITH COMPARISON SITES

This interview accompanies the Annual Perinatal Systems Survey that your hospital is completing. The purpose of this interview is to collect more in-depth information on structural and environmental factors that may influence preterm birth rates in your hospital catchment area.

1. What changes in hospital policy, if any, have been implemented during the past 12 months concerning
   (a) Elective induction
   (b) Elective primary cesarean delivery or cesarean delivery on maternal request
   (c) Newborn admission to the NICU
   (d) Smoking on the hospital campus
   (e) Provision of educational materials to women of child-bearing age
   (f) Provision of progesterone to prevent recurrent preterm birth
   (g) Other policies that may affect preterm birth rates

2. If any changes have taken place, per Question #1:
   (a) When did the changes occur?
   (b) What were the reasons for these changes?
   (c) What challenges have been faced in implementing the policy changes, and how have they been addressed?
   (d) What successes related to the policy changes should be noted?

3. What changes in patient safety/quality improvement program strategies or protocols, if any, have been implemented in the past 12 months concerning
   (a) Elective induction prior to 39 weeks
   (b) Elective primary cesarean delivery prior to 39 weeks

4. If any changes have taken place, per Question #3:
   (a) What were the reasons for these changes?
   (b) What challenges have been faced in implementing the policy changes, and how have they been addressed?
   (c) What successes related to the policy changes should be noted?

5. What changes, if any, have taken place in the past 12 months with respect to the availability of the following services (e.g., classes, medical care, social services, etc.) for pregnant women in your hospital's catchment area? (Please describe the nature, causes, and effects of these changes.)
   (a) Prenatal classes
   (b) Smoking cessation
   (c) Dental care/oral health
   (d) Domestic violence
   (e) Substance abuse
   (f) Weight management
   (g) Stress management
   (h) Other

6. What other factors, if any, have changed in the past 12 months that are affecting or will affect perinatal outcomes? Please describe. For example:
   (a) Changes in national, state, or local legislation/policies (e.g., concerning insurance coverage, reimbursable medical services, prosecution of substance-using women, smoking in public places, etc.).
   (b) Changes in funding for implementation or enforcement of relevant legislation/policies.
   (c) Changes in the availability of health-care services outside your hospital (e.g., closing or opening of other clinics or hospitals, expansion or reduction of their hours or services, etc.).
   (d) Initiation or conclusion of community-wide or targeted health prevention or promotion campaigns (e.g., media campaigns).
   (e) Other factors not discussed already in the interview.

7. Please describe any media coverage over the past 12 months on the *HBWW* initiative (e.g., timing, medium, message, target audience) that you are aware of in your hospital catchment area.

8. To the best of your knowledge, have your perinatal providers or patients participated in any *HBWW* program components or used any *HBWW* resources (e.g., used the prematurity.org Web site; attended a *HBWW* presentation at the MOD Prematurity Summit, at KPA, etc.)? If so, please describe this participation or use.

9. From 2007 through the present, what factors do you feel had the greatest impact on preterm birth rates in your hospital's catchment area? Have the rates of preterm birth gone up, stayed the same, or gone down?

ANNUAL ENVIRONMENT AND POLICY INTERVIEW WITH IS LOCAL HEALTH DEPARTMENTS

The purpose of this interview is to collect information on (1) structural and environmental factors that may influence preterm birth rates in your area, and (2) your perceptions of the impact of Healthy Babies are Worth the Wait® (*HBWW*).

1. What changes in *local or state Department for Public Health services*, if any, have taken place during the past 12 months that may be impacting perinatal outcomes in your area? These services may pertain to
   (a) Prenatal education/care
   (b) Smoking cessation
   (c) Dental care/oral health
   (d) Domestic violence
   (e) Substance abuse
   (f) Nutrition or weight management
   (g) Stress management
   (h) Other issues
   For each change, please indicate when it occurred, why it occurred, and how you think it may impact perinatal outcomes in your area.

2. Are there any changes on the horizon with respect to local or state Department for Public Health services that may impact perinatal outcomes in your area? If so, please describe.

3. What changes, if any, have taken place in the past 12 months with respect to the availability of *other services for pregnant women* (e.g., classes, medical care, social services, etc.) offered by community-based organizations, churches, private clinics, and other entities in your area?
   For each change, please indicate when it occurred, why it occurred, and how you think it may impact perinatal outcomes in your area.

4. Are there any changes on the horizon, with respect to these other services for pregnant women in the community? If so, please describe.

5. What *other factors*, if any, have changed in the past 12 months that may be impacting perinatal outcomes in your area? Please describe. For example:
   (a) Changes in state or local legislation, regulations, or policies (e.g., concerning insurance coverage, reimbursable medical services, prosecution of substance-using women, smoking in public places, etc.).
   (b) Changes in funding for implementation or enforcement of relevant legislation/policies.
   (c) Changes in funding for public health staff and programs.
   (d) Initiation or conclusion of community-wide or targeted health prevention or promotion campaigns (e.g., media campaigns).
   For each change, please indicate when it occurred, why it occurred, and how you think it may impact perinatal outcomes in your area.

6. Are there any changes on the horizon, with respect to factors such as those listed in (5) that may impact perinatal outcomes in your area? If so, please describe.

7. Please describe any media coverage of *HBWW* over the past 12 months that you are aware of in your area (e.g., timing, medium, message, target audience).

8. How would you characterize the collaboration with the *HBWW* hospital over the 3 years? What were the most successful or impactful aspects? What could have been done to make this collaboration even better?

9. (a) To date, what do you perceive to be the most valuable or impactful aspects of *HBWW*?
   (b) What valuable or impactful changes implemented because of *HBWW* will you sustain?

10. To date, what do you believe to have been the least impactful aspects of *HBWW*?

APPENDIX H
HBWW NEWSLETTER

HEALTHY BABIES ARE
WORTH THE WAIT
NEWSLETTER

Mara Burney, Editor
mburney@marchofdimes.com

Healthy Babies
ARE WORTH THE WAIT℠
Volume I, Issue I

ISSUE I   SUMMER, 2007

HBWW Comparison Sites:
- Western Baptist Hospital (Paducah)
- Norton Hospital Downtown (Louisville)
- Lake Cumberland Regional Hospital (Somerset)

Inside the *Healthy Babies Are Worth the Wait* Launch

As the first state in the nation to implement *Healthy Babies Are Worth the Wait* (HBWW), Kentucky is not only enhancing the health of its own citizens, but also paving the way for an exciting new preterm birth prevention strategy that will ultimately benefit families across the country. This was the message that politicians, public health officials, and medical professionals joined in announcing at the HBWW launch, which took place during the Kentucky Public Health Association meeting in Louisville on March 27.

Approximately 150 people were in attendance at the news conference held at the Executive West Hotel in Louisville, KY on Tuesday, March 27, 2007 to launch the Initiative. Guests included the Governor of Kentucky, Kentucky Public Health Association, as well as March of Dimes and

Johnson & Johnson Pediatric Institute representatives, staff and board members.

The launch event received extensive media attention, including placements in The Lexington Herald Leader (front page story) The Courier-Journal of Louisville (front page of the Metro Section), The Paducah Sun, and The New Jersey Herald. Television coverage included stations in Louisville, Lexington, and Paducah. The launch was also covered on 192 web sites, including CNN.com, Yahoo.com, USAToday.com, Reuters.com, Latimes.com, Kentucky.com, AOL.com, the Kentucky Cabinet for Health and Family Services web site, the Kentucky Commonwealth News Center, the Johnson & Johnson Pediatric Institute web site, the March of Dimes web site, and the SUNY Upstate web site.

Congratulations and thank you to everyone who helped make the launch a success!

Look inside for more photos!

Meet the Intervention Sites: Three hospitals, three models of care

King's Daughters Medical Center, University of Kentucky, and Trover Clinic/Regional Medical Center Hopkins County are the first three hospitals selected to receive the HBWW interventions.

In the East... King's Daughters Medical Center (KDMC) is a not-for-profit, 385-bed regional referral center, covering a 150-mile radius. The majority of the approximately 1700 deliveries that occur each year here are to women cared for by private doctors.

In the Center... Established in 1957, UK Healthcare is an academic center that includes medical, nursing,

public health, health sciences, pharmacy, and dentistry patient care activities. Most of UK's approximately 1900 deliveries per year are to women cared for in a clinic setting. However, many of UK's Hispanic moms-to-be participate in group prenatal care.

In the West... Regional Medical Center of Hopkins Count ("Trover Clinic") serves a 12-county area in Western Kentucky of approximately 285,000 people. The majority of Trover's approximately 1000 deliveries per year are to patients participating in group prenatal care (see page 2). *Preliminary delivery data for 2006 courtesy of Tracey Jewell, KY DPH*

Photos: Governor Ernie Fletcher speaks at the HBWW Launch (right) Dr. James Ferguson of University of Kentucky and Jennifer Sparks of King's Daughters Medical Center (top) and LeAnn Todd of Trover Clinic (left) speak on behalf of their sites, SUNY Upstate Medical Center President Dr. David Smith, March of Dimes President Dr. Jennifer Howse, Governor Ernie Fletcher, Johnson & Johnson Pediatric Institute Advisor Dr. Steven Shelov, KPHA President Shawn Crabtree together at the event (bottom).

Spotlight: Centering at Trover Clinic

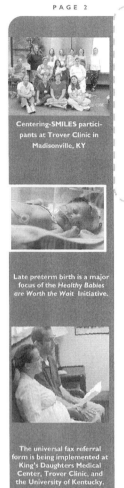

Centering-SMILES participants at Trover Clinic in Madisonville, KY

Late preterm birth is a major focus of the *Healthy Babies are Worth the Wait* Initiative.

The universal fax referral form is being implemented at King's Daughters Medical Center, Trover Clinic, and the University of Kentucky.

By LeAnn Todd

Trover Health System is a multi-specialty group practice located in the heart of western Kentucky. Trover's Center for Women's Health is a diverse practice focusing on care of women throughout the lifespan, including preconception, prenatal care, adolescent and adult gynecology, and postmenopausal care.

A core service of the prenatal care program at Trover Health System is the **CenteringPregnancy** program. Centering is comprehensive prenatal care, which brings the prenatal visit out of the "exam room" setting and into a group setting. Groups consist of about 10 women with similar due dates, and are led by a Certified Nurse-Midwife (CNM). At the start of each meeting every woman gets a few minutes alone with the CNM to listen to the baby's heartbeat and get a general "belly check." The next step in Centering is "the circle". The women of the group meet, sitting in a circle, to ask questions, discuss concerns, and share experiences.

Topics discussed in each group may be suggested by the CNM based on the current point in pregnancy, or may come directly from the group according to their individual needs and concerns. Expert advice comes to the group from the CNM, nurses, pediatricians, anesthetists, and from the wisdom of the women themselves. Centering also provides comprehensive dental screening and treatment for oral infections to promote healthier pregnancies.

The providers and staff are eager to improve pregnancy outcomes by participating in the Kentucky Prematurity Prevention Initiative, because *Healthy Babies Are Worth the Wait!*

New Information About Late Preterm Birth

By Karla Damus

In 2006, new information about the risks of being born even a few weeks early was published in *Seminars in Perinatology* and in *Clinics in Perinatology*. Much of this information came from presentations by expert clinicians and scientists at the 2005 invitational conference led by NICHD, with representatives from ACOG, AAP, AWHONN, and March of Dimes.

The data and studies identified greater short and long term morbidities for babies born late preterm (34-36 weeks-previously referred to as "near term") compared to babies born at term (37-41 weeks).

In the US and KY, late preterm infants drive the overall increases in preterm birth rates. They are the majority (71%+) of all preterm births and of all NICU admissions, and utilize most of the respiratory support in the NICU. They are also at increased risk of infant mortality, SIDS, re-hospitalization, ADHD, and developmental delay compared to term infants. Between 1994 and 2004, the rate of late preterm births increased from 8% to 10.2% in KY, which is twice as fast as the nation's rate.

This is why *Healthy Babies are Worth the Wait* is targeting a reduction in late preterm birth rates to help reduce the overall rates of preterm birth in the state. It is also why the Initiative sponsored Dr. Raju to deliver the keynote at the KPA and also to then give grand rounds at UK the following day. Bringing national experts like Dr. Raju to KY to provide continuing education is a priority of the Initiative, so that providers can have the latest information to inform the high quality care they deliver to families in Kentucky.

Making Connections: The HBWW Universal Fax Referral Form

By Katrina Thompson

Healthy Babies are Worth the Wait has developed a new way to connect patients at intervention hospitals with the health department services they need: a universal fax referral form. The referral form allows providers to refer women to one source by simply checking off the needed services which are then followed up by the Health Department staff and/or other appropriate providers. The available services include, but are not limited to, programs such as breastfeeding classes, childbirth preparation classes, WIC, HANDS (home visitation program), substance abuse programs, domestic violence services, dental services, prenatal diabetes counseling and smoking cessation services including Kentucky's smoking Quit Line for pregnant women, group therapy, and individual counseling.

By simply supplying one signature the patient can rest assured that advocates are working on their behalf to ensure that any services needed are further addressed, and that their referral form remains confidential in accordance with HIPAA guidelines. The patient can request with a simple checked box that their participation, including progress reports, be shared with their health care provider, further enhancing the patients overall health and wellness.

ISSUE 1   PAGE 3

Smoking Cessation in Pregnancy Trainings

by Ruth Ann Shepherd

In April 2007, the Kentucky Department for Public Health's Tobacco Cessation and Control Program hosted a week of meetings and trainings entitled "Women and Tobacco: Consequences and Solutions". One of the featured speakers for this event was Dr. Cathy Melvin, a leading expert and researcher on smoking during pregnancy and Chair of the National Partnership to Help Pregnant Smokers Quit. Dr. Melvin is currently an Associate Professor at the UNC Chapel Hill School of Public Health and has had a distinguished career in maternal and child health, including such positions as Research Fellow and Director of Child Health Services at the Sheps Center for Health Services Research at UNC, Director for Smoke Free Families, Senior Scientist, Division of Reproductive Health, National Center for Chronic Disease Prevention and Health Promotion at CDC.

Dr. Melvin presented an update on treating tobacco dependence in pregnant women. Over 100 people participated, including representatives from the Intervention Sites. Dr. Melvin reviewed the consequences and costs of smoking in pregnancy. Smoking almost doubles the cost of treating health problems during pregnancy. Most women these days know smoking during pregnancy is bad; what they need is help and encouragement, not lectures on the bad effects of smoking. Pregnant women who smoke don't believe their physicians can or will help them quit smoking. Many fear they will only get a judgmental attitude and lecture if they try and discuss smoking with

their providers. Some feel the issue is "not important enough" for the provider to take the time to discuss.

In some ways, they may correct. Simple advice from the provider to stop smoking is not effective, but brief counseling is. The most commonly used evidence-based program is the 5A's. It is proven to be effective, but very few systems of care for pregnant women have implemented this technique.

About 18-25% of women smokers stop smoking on their own when they learn they are pregnant. Consistent implementation of the 5A's, with patient self-help materials, can increase the rate of quitting by 30-70%. Women are more likely to quit smoking during pregnancy than any other time in their lives, so intervention to assist them during this time is critical. Referral to a Quit Line also improves quit rates, and works best in combination with individual or group counseling. The national quit line has staff trained specifically to deal with pregnant smokers.

The next day, Dr. Melvin traveled to Ashland for a workshop at Kings Daughter's Medical Center, where about 20 people who interact with pregnant women in different venues attended. Dr. Melvin trained the group on the 5A's using an interactive CD developed by Dartmouth Interactive Media Lab and available through ACOG. The workshop included discussion of innovative ways to assist and encourage pregnant smokers to quit. Some of ideas from the group included:

- Standing orders for nicotine replacement or diversions for smoking pregnant women when they are hospitalized (Kings Daughters

was at the time implementing a "smoke free campus" policy)
- Diversions or nicotine gum/patch to suggest to the patient for that immediate post-partum need for a cigarette
- Referral to counseling and/or the quit line at every opportunity
- Educate providers on availability of smoking cessation counseling, individual or group, in their community (Ask, Assist & Refer)
- Have tobacco cessation counselors go to private OB offices one day a week to train staff and be available to counsel patients there
- Materials for waiting rooms both in doctor's offices and in the hospital
- Group classes for smoking cessation tailored for pregnant women
- Go to where the women are – other community groups, gatherings (churches, groceries, beauticians, etc). Women often respond better to community and worksite interventions.
- Consider incentives (e.g., a free photo of the baby) for mothers who stay off cigarettes during the pregnancy (monitor conitine levels)
- Enlist the help of a friend or family member to support the mother who is trying to quit.

Dr. Melvin discussed this as an opportunity for a Quality Improvement project that would be applicable for a hospital or health department – to map the processes where pregnant women are served every day, and then determine how she could get repeated encouragement, consistent messages, and help for quitting from everyone.

"Most women these days know smoking during pregnancy is bad; what they need is help and encouragement, not lectures on the bad effects of smoking. "

Additional Resources:

- 5A's Training: http://iml.dartmouth.edu/education/cme/Smoking

- www.helppregnantsmokersquit.org

- www.smokefreefamilies.org

- Kentucky Quit Line 1-800-784-4667

- Your Local Health Department

MOD-JJPI Grand Rounds Program

Shop online, get driving directions online, and now, attend grand rounds online! This year, the March of Dimes-Johnson & Johnson Pediatric Institute Grand Rounds Program (GRP) is adding a web-based component to supplement its traditional hospital-based grand rounds. This means that perinatal providers across Kentucky can now enjoy nationally-renowned speakers on topics related to preterm labor and birth (not to

mention free continuing education credits!) from the comfort of their own homes or offices. In addition, the webcasts will be archived and available for up to one year following the live presentation.

June's presentation featured Dr. Ruth Ann Shepherd speaking on Late Preterm Birth: Rates, Risks, and Impact. Upcoming presentations are planned for August, October, and November, and will cover promising

interventions to prevent premature birth (such as progesterone and smoking cessation), pre– and interconception health and management, and multiples and infertility management. You are welcome to participate even if you do not need CE credits. *To view the archived late PTB webcast, go to cdnetwork.org and click on Pregnancy/OB-GYN in the Webcast Library. More information: grandrounds@marchofdimes.com.*

Dr. Ruth Ann Shepherd presented the first MOD-JJPI webcast on June 14

PAGE 4

" Can we reverse the rising rate of premature birth? Here in Kentucky, this is no longer a rhetorical question: over the next three years, we intend to find out. Our goal is to reduce singleton premature birth by 15% among the 6,000 women we will reach through this initiative. That would mean 100 premature births would be prevented – 100 more babies would go full term. "
-Dr. Jennifer Howse

"Pediatricians share responsibility with our OB-GYN colleagues to do everything possible to ensure that babies get a healthy start in life, because all of us who work in the public health arena have witnessed the explosion of premature births firsthand -- with the growing number of preemies in our practices, our clinics, and our hospitals. I can say, as a physician who has cared for these babies and been with the parents, even one baby born too soon is one too many if it's *your* child."
-Dr. Steven Shelov

For information on the Initiative, contact:

March of Dimes (Kentucky)
Katrina Thompson (859)-246-0004
March of Dimes (National)
Mara Burney (914) 997-4287
Johnson & Johnson Pediatric Institute, L.L.C.
Bonnie Petrauksas (732) 524-6172
Kentucky Department for Public Health
Dr. Ruth Ann Shepherd (502) 564-4830

• KD

King's Daughters Medical Center
Jennifer Sparks (606) 327-4243
Ashland-Boyd Health Department
Maria Hardy (606) 329-9444

• TC

Trover Clinic (Women's Center)
LeAnn Todd (270) 326-3921
Hopkins County Health Department
Jack Morris (270) 821-5242

• UK

University of Kentucky Dept. of OB-GYN
Dr. James Ferguson (859) 257-2323
Lexington-Fayette Health Department
Dr. Melinda Rowe (859) 288-2846

Each of the three intervention sites expressed their enthusiasm for the project and their optimism for positive outcomes. In addition, Dr. Ferguson acknowledged the commitment and leadership of Dr. Rowe and her staff from the Lexington-Fayette Health Department.

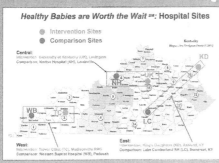

Healthy Babies are Worth the Wait ℠: Hospital Sites

● Intervention Sites
● Comparison Sites

The HBWW Initiative includes 3 intervention Sites and 3 Comparison Sites spread across the state of Kentucky.

Left to right: Initiative team members Dr. Ruth Ann Shepherd, Dr. Diane Ashton, Bonnie Petrauskas, Katrina Thompson, Mara Burney, and Dr. Karla Damus

Dr. Julie Solomon of Sociometrics, Inc. in Los Altos, CA is the independent evaluator for Healthy Babies Are Worth the Wait.

Healthy Babies Are Worth the Wait ᵉᵐ is helping Kentucky's babies get the best possible start in life. Working with health care providers and community partners, the program helps ensure that moms-to-be have the care and information they need to maintain healthy, full-term pregnancies. The goal of the Initiative is a 15% reduction in preterm births in the Intervention Sites over the 3 year intervention period.

Healthy Babies Are Worth the Wait ᵉᵐ is an initiative of the Prematurity Prevention Partnership: the March of Dimes, Johnson & Johnson Pediatric Institute, and the Kentucky Department for Public Health.

For further information about the Initiative please call 859-246-0004 or visit prematurityprevention.org.

Lakeside with *Healthy Babies Are Worth the Wait*
The 2007 Kentucky Perinatal Association Meeting
A panel of HBWW speakers presented at this year's KPA at the Lake Cumberland Resort. Topics included an overview of the HBWW Initiative, progesterone to prevent recurrent PTB, public health collaborations in KY, and HBWW surveys. The MOD-JJPI GRP also sponsored a talk by Dr. Tonse Raju on the problem of late PTB. Slides from the presentations are available on the KPA website at the address below.

http://www.kentuckyperinatal.com/June_2007_Meeting_Recap.html

Top left: Record-breaking attendance at this year's KPA. Representatives from each Intervention Site attended the KPA on a HBWW scholarship. Top right: Dr. Tonse Raju of NICHD gives the keynote luncheon address.

Dr. Damus bares it all at the Kentucky Perinatal Association meeting in Jamestown on June 4

Late PTB Quiz

1. "Late preterm birth" refers to...?

A. 32-35 weeks

B. 34-36 weeks

C. 34-37 weeks

D. Not getting to L&D on time to perform a delivery

2. At 35 weeks of gestation, the fetal brain is what percent of the weight it will be at term?

A. 85%

B. 75%

C. 66%

Answers: 1. B 34-36 weeks 2. C 66%

Upcoming Events

• August 11 Breastfeeding Event at Babies 'R' US in Lexington

• Sept. 17- 2nd Annual Site Council Meeting in Frankfort, KY

• September 27th Kentucky Folic Acid Partnership Celebration at Buffalo Trace (Frankfort)

• November 13 Kentucky Prematurity Summit in Louisville, KY (will be preceded by HBWW meeting on November 12... details to follow)

REFERENCES

[1] Martin J, Hamilton B, Sutton P, Ventura S, Menacker F, Kirmeyer S, et al. Births: final data for 2006. Natl Vital Stat Rep 2009;57(7):1–102.

[2] Committee on Obstetric Practice. ACOG Committee Opinion No. 404 April 2008. Late preterm infants. Obstet Gynecol 2008;111(4):1029–32. http://dx.doi.org/10.1097/AOG.0b013e31817327d0.

[3] Institute of Medicine (IOM) (US) Committee on Understanding Premature Birth and Assuring Healthy Outcomes. In: Behrman RE, Butler AS, editors. Preterm birth: causes, consequences, and prevention. Washington (DC): National Academies Press (US); 2007.

[4] Mathews TJ, MacDorman MF. Infant mortality statistics from the 2006 period linked birth/infant death data set. Natl Vital Stat Rep 2010;58(17):1–31.

[5] Ramachandrappa A, Jain L. Health issues of the late preterm infant. Pediatr Clin North Am 2009;56(3):565–77. http://dx.doi.org/10.1016/j.pcl.2009.03.009.

[6] Engle WA, Tomashek KM, Wallman C, Committee on Fetus and Newborn, American Academy of Pediatrics. "Late-preterm" infants: a population at risk. Pediatrics 2007;120(6): 1390–401. http://dx.doi.org/10.1542/peds.2007-2952.

[7] Tomashek KM, Shapiro-Mendoza CK, Davidoff MJ, Petrini JR. Differences in mortality between late-preterm and term singleton infants in the United States, 1995–2002. J Pediatr 2007;151(5):450–6, 456.e1. http://dx.doi.org/10.1016/j.jpeds.2007.05.002.

[8] Morse SB, Zheng H, Tang Y, Roth J. Early school-age outcomes of late preterm infants. Pediatrics 2009;123(4):e622–9. http://dx.doi.org/10.1542/peds.2008-1405.

[9] Green NS, Damus K, Simpson JL, March of Dimes Scientific Advisory Committee on Prematurity. Research agenda for preterm birth: recommendations from the March of Dimes. Am J Obstet Gynecol 2005;193(3 Pt 1):626–35. http://dx.doi.org/10.1016/j.ajog.2005.02.106.

[10] NIH state-of-the-science conference statement on cesarean delivery on maternal request. NIH Consens State Sci Statements March 27–29, 2006;23(1):1–29.

[11] Tita AT, Landon MB, Spong CY, Lai Y, Leveno KJ, Varner MW, et al. Eunice Kennedy Shriver NICHD Maternal-Fetal Medicine Units Network. Timing of elective repeat cesarean delivery at term and neonatal outcomes. N Engl J Med 2009;360(2):111–20. http://dx.doi.org/10.1056/NEJMoa0803267.

[12] American College of Obstetricians and Gynecologists. ACOG Practice Bulletin No. 107: induction of labor. Obstet Gynecol 2009;114(2 Pt 1):386–97. http://dx.doi.org/10.1097/AOG.0b013e3181b48ef5.

[13] American College of Obstetricians and Gynecologists. ACOG Practice Bulletin No. 115: Vaginal birth after previous cesarean delivery. Obstet Gynecol 2010;116(2 Pt 1): 450–63. http://dx.doi.org/10.1097/AOG.0b013e3181eeb251.

[14] Engle WA, Kominiarek MA. Late preterm infants, early term infants, and timing of elective deliveries. Clin Perinatol 2008;35(2):325–41, vi. http://dx.doi.org/10.1016/j.clp.2008.03.003.

[15] Bettegowda VR, Dias T, Davidoff MJ, et al. The relationship between cesarean delivery and gestational age among US singleton births. Clin Perinatol 2008;35(2):309–23, v-vi. http://dx.doi.org/10.1016/j.clp.2008.03.002.

[16] Last JM, editor. A dictionary of epidemiology. 3rd ed. New York (NY): Oxford University Press; 1995.

[17] Morgenstern H, Thomas D. Principles of study design in environmental epidemiology. Environ Health Perspect 1993;101(Suppl. 4):23–38.

[18] Martin JA, Kirmeyer S, Osterman M, Shepherd RA. Born a bit too early: recent trends in late preterm births. NCHS Data Brief 2009;24:1–8.

[19] Savitz DA, Murnane P. Behavioral influences on preterm birth: a review. Epidemiol (Cambridge, MA) 2010;21(3):291–9. http://dx.doi.org/10.1097/EDE.0b013e3181d3ca63.

[20] Bukowski R, Malone FD, Porter FT, et al. Preconceptional folate supplementation and the risk of spontaneous preterm birth: a cohort study. PLoS Med 2009;6(5):e1000061. http://dx.doi.org/10.1371/journal.pmed.1000061.

[21] Jain L, Raju T, editors. Clinics in Perinatology. Late preterm pregnancy and the newborn, vol. 33, no. 4. Philadelphia (PA): W.B. Saunders Company; December 2006.

[22] American College of Obstetricians and Gynecologists. ACOG Practice Bulletin no. 130: prediction and prevention of preterm birth. Obstet Gynecol 2012;120(4):964–73. http://dx.doi.org/10.1097/AOG.0b013e3182723b1b.

[23] Ickovics JR, Kershaw TS, Westdahl C, Magriples U, Massey Z, Reynolds H, et al. Group prenatal care and perinatal outcomes: a randomized controlled trial. Obstet Gynecol 2007;110(2 Pt 1):330–9. http://dx.doi.org/10.1097/01.AOG.0000275284.24298.23.

[24] Rising SS, Kennedy HP, Klima CS. Redesigning prenatal care through CenteringPregnancy. J Midwifery Women's Health 2004;49(5):398–404. http://dx.doi.org/10.1016/j.jmwh.2004.04.018.

[25] Skelton J, Mullins R, Langston LT, Womack S, Ebersole JL, Rising SS, et al. CenteringPregnancy Smiles: Implementation of a small group prenatal care model with oral health. J Health Care Poor Underserved 2009;20(2):545–53. http://dx.doi.org/10.1353/hpu.0.0138.

[26] Moore S, Ide M, Coward PY, et al. A prospective study to investigate the relationship between periodontal disease and adverse pregnancy outcome. Br Dent J 2004;197(5): 251–8; discussion 247. http://dx.doi.org/10.1038/sj.bdj.4811620.

[27] American College of Obstetricians and Gynecologists, Women's Health Care Physicians, & Committee on Health Care for Underserved Women. Committee Opinion No. 569: oral health care during pregnancy and through the lifespan. Obstet Gynecol 2013; 122(2 Pt 1):417–22. http://dx.doi.org/10.1097/01.AOG.0000433007.16843.10.

[28] Jeffcoat MK, Geurs NC, Reddy MS, et al. Periodontal infection and preterm birth: results of a prospective study. J Am Dent Assoc (1939) 2001;132(7):875–80.

[29] Offenbacher S, Boggess KA, Murtha AP, et al. Progressive periodontal disease and risk of very preterm delivery. Obstet Gynecol 2006;107(1):29–36. http://dx.doi.org/10.1097/01.AOG.0000190212.87012.96.

[30] Klebanoff M, Searle K. The role of inflammation in preterm birth–focus on periodontitis. BJOG: Int J Obstet Gynaecol 2006;113(Suppl. 3):43–5. http://dx.doi.org/10.1111/j.1471-0528.2006.01121.x.

[31] Barker DJP, Hall AJ. Practical epidemiology. Churchill Livingstone; 1991.

[32] Ness A, Dias T, Damus K, Burd I, Berghella V. Impact of the recent randomized trials on the use of progesterone to prevent preterm birth: a 2005 follow-up survey. Am J Obstet Gynecol 2006;195(4):1174–9. http://dx.doi.org/10.1016/j.ajog.2006.06.034.

[33] Bernstein P, Berck D, Burgess T, et al. Preventing preterm birth: the role of 17P. Albany (NY): American College of Obstetricians and Gynecologists, District II; 2009.

[34] Martin JA, Hamilton BE, Sutton PD, et al. Births: final data for 2007. Natl Vital Stat Rep 2010;58(24):1–85.

[35] Martin JA, Hamilton BE, Ventura SJ, et al. Births: final data for 2009. Natl Vital Stat Rep 2011;60(1):1–70.

[36] Leveno KJ, McIntire DD, Bloom SL, et al. Decreased preterm births in an inner-city public hospital. Am J Obstet Gynecol 2009;113(3):578–84.

INDEX

Note: Page numbers followed by "f" and "t" indicate figures and tables respectively.

Printed in the United States
By Bookmasters

Printed in the United States
By Bookmasters